Carbohydrate Chemistry

Monosaccharides and Their Oligomers

Carbohydrate Chemistry

Monosaccharides and Their Oligomers

Hassan S. El Khadem

The American University
Washington D.C.

ACADEMIC PRESS, INC.
Harcourt Brace Jovanovich, Publishers
San Diego New York Berkeley Boston
London Sydney Tokyo Toronto

ACADEMIC PRESS, INC.
1250 Sixth Avenue
San Diego, California 92101

United Kingdom Edition published by
ACADEMIC PRESS INC. (LONDON) LTD.
24-28 Oval Road, London NW1 7DX

Library of Congress Cataloging-in-Publication Data

El Khadem, Hassan Saad, Date
Carbohydrate chemistry.

Bibliography: p.
Includes index.
1. Monosaccharides. 2. Carbohydrates. I. Title.
QD321.E48 1988 547.7′813 87-47760
ISBN 0-12-236870-3 (alk. paper)

PRINTED IN THE UNITED STATES OF AMERICA
88 89 90 91 9 8 7 6 5 4 3 2 1

Contents

II. Oligomeric Saccharides: Oligosaccharides and Nucleotides

List of Tables

Preface

Carbohydrate Chemistry: Monosaccharides and Their Oligomers was designed as a classroom textbook for college carbohydrate chemistry courses. It is primarily intended for use by undergraduate and graduate students enrolled in chemistry, biochemistry, and pre-med curricula. It is also intended for students in the colleges of arts and sciences, pharmacy, agriculture, and medicine, who are engaged in research in the fields of carbohydrates and natural product chemistry. Its format fills a gap between large, multivolume reference books designed mainly for research and elementary, sometimes superficial books.

Review articles dealing with topics discussed in the various chapters are listed in chapter order in the bibliography at the end of the book. In addition, references for a small number of recent articles, not found in these reviews, appear throughout the text. Some problems on NMR spectroscopy and other topics have also been added to Chapters 3–7. Answers to the problems are in the appendix at the back of the book.

The book is divided into two parts. The first, "Monosaccharides," deals with monomeric carbohydrates, whereas oligosaccharides and oligonucleotides are discussed in the second. Chapter 1 is an introduction that outlines the importance of carbohydrates as a major group of naturally occurring compounds. After a short historical section on the discovery of sugars, the classification of carbohydrates is discussed in detail.

The four chapters that follow deal with the chemistry of monosaccharides. Thus, Chapter 2 discusses the determination of the structure, configuration, and conformation of monosaccharides. Although the elucidation of the structure of monosaccharides was achieved at a time when sophisticated instrumental techniques were not available, infrared and nuclear magnetic resonance (NMR) spectroscopy, as well as mass spectrometry, are extensively discussed in this chapter. Nomenclature is discussed next, in a short chapter that illustrates the proper use of the rules adopted by the International Committee on Carbohydrate Nomenclature and by *Chemical Abstracts*.

The two longest chapters in Part I are Chapter 4, "Physical Properties Used in Structure Elucidation," and Chapter 5, "Reactions of Monosaccharides." The first deals with spectroscopic methods, such as NMR (^{1}H, ^{13}C, and ^{15}N) and molecular and electronic spectroscopy, as well as mass spectrometry. These, together with the optical properties of monosaccharides (optical rotation, optical rotatory dispersion, and circular dichroism), are extensively used today in structure determinations in the field of carbohydrates. Chapter 5 deals with the reactions of monosaccharides, starting with addition reactions of the carbonyl group. These are followed by nucleophilic substitution reactions at the anomeric carbon atom and at the less reactive nonanomeric carbon atoms. Oxidations and reductions of monosaccharides are then discussed, followed by a section on how to plan retrosynthetic schemes.

Part II starts with Chapter 6 on the methods used for the structure elucidation of oligosaccharides. These include wet chemical methods suitable for the study of oligomers available in gram quantities and nondestructive physical methods capable of handling milligram quantities of rare oligomers. The synthesis of oligosaccharides and their chemical modification is treated in Chapter 7, which includes some novel approaches to the synthesis of these oligomers.

The author is deeply indebted to R. S. Tipson, who read the entire manuscript and made several useful suggestions. Thanks are also due to Derek Horton and Stephen Hanessian, who read a second draft of the manuscript and made valuable suggestions. In addition, D. L. Swartz offered suggestions and helped with the operation of a word processor. Many figures and diagrams were made available by authors for use in the text; for example, photographs of neutron diffraction analyses were presented by G. A. Jeffrey, and of NMR spectra by B. Coxon and A. C. Allen. Their help and that of other friends is greatly appreciated.

1

Carbohydrates

I. HISTORICAL BACKGROUND

The origins of carbohydrate chemistry can be traced back to the civilizations of antiquity. Thus, for example, the manufacture of beer and wine by alcoholic fermentation of grain starch and grape sugar is well documented on the walls of ancient Egyptian tombs. The isolation of cellulose fibers by retting flax was widely used by the civilizations of the Far East and the Near East and was introduced by the Greeks to Europe. It is also known that gums and resins were valued commodities at the dawn of the Christian era.

The isolation of sucrose from the juice of sugarcane marks an important milestone in sugar chemistry. In the Far East, where this plant grew, sugar was isolated as a yellowish syrupy concentrate which crystallized, on standing, into a brown mass, and a number of Chinese recipes dating from the fourth century describe in detail how the sugarcane juice was concentrated. It is interesting that in Sanskrit the word *sugar* means *sweet sand,* which aptly describes the properties of crushed raw sugar. Significant progress in the manufacture of sugar occurred after the French Revolution, when Europe, under blockade, had to rely on the sugar beet for manufacture of this valued commodity. Improved methods of isolation and refining were developed, including treating the syrup with lime to precipitate the calcium complex, regenerating the sugar with sulfur dioxide, and decolorizing it with animal charcoal.

Another important industrial process in use today that may be traced to Far Eastern origins is the manufacture of paper from wood pulp. This was achieved around A.D. 600 in China and constituted a marked improvement over the then-available Egyptian papyri. The paper produced was, however, brittle and tore when folded, because the lignin had not been removed. The art of papermaking was brought by Marco Polo to Europe, where methods for removing lignin from the pulp were devised, first using alkali and then sulfite.

II. IMPORTANCE OF CARBOHYDRATES

In addition to the manufacture of sucrose and paper, which are today important industries, carbohydrates play a major role in a number of other industries. These include (a) the food industry, which uses huge amounts of starch in various degrees of purity in the manufacture of baked goods and pastas, of gums in food processing, and of mono- and oligosaccharides as sweeteners, and employs them in fermentation to make beer and wine; (b) the textile industry, which, despite the advent of synthetics, is still dependent to a large extent on cellulose; (c) the pharmaceutical industry, particularly in the areas of antibiotics, intravenous solutions, and vitamin C; and (d) the chemical industry, which produces and markets several pure sugars and their derivatives.

Carbohydrates also play a key role in the process of life; thus, the *master molecule* DNA is a polymer made up of repeating units composed of four nucleotides of 2-deoxy-D-*erythro*-pentofuranose ("2-deoxy-D-ribose"), whose sequence constitutes the coded template responsible for replication and transcription. The science of molecular biology is based on the properties of the polymers of these nucleotides. Saccharide derivatives also form part of many vital enzyme systems, specifically as coenzymes. They are also responsible for cell recognition and are therefore of great importance in immunology.

The best known use of carbohydrates is undoubtedly in nutrition, as members of a major food type which is metabolized to produce energy. Although the average percentage of carbohydrates consumed by humans in comparison to other food types differs from country to country, and figures are often inaccurate or unavailable, the value for the world as a whole has been estimated at over 80%. The exothermic reactions that produce energy in the cell are the outcome of a number of complex enzyme cycles that originate with hexoses and end up with one-, two-, or three-carbon units. The energy released from these reactions is stored in the cell in the form of a key sugar intermediate, adenosine triphosphate

(ATP), and released when needed by its conversion into the di- or monophosphate. Finally, it should be realized that carbohydrates are the most abundant organic components of plants (more than 50% of the dry weight) and therefore constitute the major part of our renewable fuels and the starting material from which most of our fossil fuels were engendered.

III. DEFINITIONS AND SCOPE

In the nineteenth century, when the empirical formulas of many organic compounds were being determined, it was discovered that all of the sugars known at the time had the formula $C_x(H_2O)_y$. They were accordingly thought to be hydrated carbons and so were called *carbohydrates*. This name was applied not only to the soluble sugars but also to polysaccharides, such as starch, that were thought, because of their formulas, to be hydrated carbons. Today's usage of the word *carbohydrate* applies to a large number of organic compounds, monomeric, oligomeric, and polymeric in nature, which do not necessarily have their hydrogen and oxygen atoms in the molecular ratio of 2 : 1 but which can be either synthesized from or hydrolyzed to monosaccharides.

Because the word *sugar* has been used to mean (a) sucrose by the general public, (b) glucose by the medical profession, and (c) mono- to oligosaccharides by chemists, it is slowly being replaced by the less ambiguous term *saccharide,* meaning sugarlike, which is usually prefixed by mono-, di-, oligo-, or poly- to designate the degree of polymerization of the specific compound under discussion. Thus, today one speaks of the chemistry of *carbohydrates* and the reactions of *monosaccharides*.

IV. CLASSIFICATION OF CARBOHYDRATES

Carbohydrates are classified, according to their degree of polymerization, into monomeric carbohydrates, which include monosaccharides and their derivatives, and polymeric carbohydrates, which comprise oligosaccharides, polysaccharides, DNA, and RNA. These polymeric carbohydrates differ in the type of bridge that links their monosaccharide units. Thus, oligosaccharides and polysaccharides are polyacetals, linked by acetal oxygen bridges, whereas DNA and RNA are poly(phosphoric esters), linked by phosphate bridges. In addition to these well-defined groups of carbohydrates, there exist a number of derivatives, for example, antibiotics, that are best studied as a separate group, because some of their members may be monomeric, whereas others are oligomeric.

A. Monosaccharides

Monosaccharides are chiral polyhydroxyalkanals or polyhydroxyalkanones which often exist in cylic hemiacetal forms. Monosaccharides are divided into two major groups according to whether their acyclic forms possess an aldehyde group or a keto group, that is, into aldoses or ketoses (glyculoses). These, in turn, are each classified, according to the number of carbon atoms in the monosaccharide chain (usually 3–10), into trioses, tetroses, pentoses, hexoses, etc. By prefixing aldo- to these names one may define more closely a group of aldoses, for example, aldopentoses, whereas for ketoses it is customary to use the ending -ulose, as in hexuloses. Finally, monosaccharides may be grouped according to the size of their rings into five-membered furanoses and six-membered pyranoses. It should be noted that, in order to form a furanose ring, four carbon atoms and one oxygen atom are needed, so only aldotetroses and higher aldoses and 2-pentuloses and higher ketoses can cyclize in this ring form. Similarly, in order to form a pyranose ring, five carbon atoms and one oxygen atom are required, so only aldopentoses and 2-hexuloses as well as their higher analogs can cyclize in this form. Ultimately, by combining the ring type with the names used above (for example, aldopentose or hexulose), such combination names as aldopentofuranoses and hexulopyranoses can be formed, which define without ambiguity the group to which a monosaccharide belongs (see Table I).

As their name denotes, monosaccharides are monomeric in nature and, unlike the oligosaccharides and polysaccharides, which will be discussed later, they cannot be depolymerized by hydrolysis to simpler sugars. Monosaccharides and oligosaccharides are soluble in water and their solutions in water are often sweet-tasting, which is why they are referred to as sugars.

B. Oligosaccharides

Oligosaccharides and polysaccharides are polyacetals which respectively have, as their names denote (from the Greek *oligos*, few; and *poly*, many), a low (DP = 2–10) or a high (DP >10) degree of polymerization. They are composed of a number of monosaccharides linked together by acetal oxygen bridges, and, upon depolymerization (hydrolysis), they yield one or more types of monosaccharide. Oligosaccharides and polysaccharides are further grouped into (a) simple (true) oligosaccharides and polysaccharides, which are oligomers and polymers of monosaccharides that yield on *complete* hydrolysis *only* monosaccharides; and (b) conjugate oligosaccharides and polysaccharides, which are oligomers and poly-

TABLE I
Monosaccharides

Aldoses				
Monosaccharide	Aldose[a]	No. of chiral C	Aldofuranose	Aldopyranose
Triose	Aldotriose	1	—	—
Tetrose	Aldotetrose	2	Tetrofuranose; aldotetrofuranose	— —
Pentose	Aldopentose	3	Pentofuranose; aldopentofuranose	Pentopyranose; aldopentopyranose
Hexose	Aldohexose	4	Hexofuranose; aldohexofuranose	Hexopyranose; aldohexopyranose
Heptose	Aldoheptose	5	Heptofuranose; aldoheptofuranose	Heptopyranose; aldoheptopyranose
Octose	Aldo-octose	6	Octofuranose; aldo-octofuranose	Octopyranose; aldo-octopyranose
Nonose	Aldononose	7	Nonofuranose; aldononofuranose	Nonopyranose; aldononopyranose
Decose	Aldodecose	8	Decofuranose; aldodecofuranose	Decopyranose; aldodecopyranose

Ketoses (Glyculoses)				
Monosaccharide	Ketose[b]	No. of chiral C	Ketofuranose	Ketopyranose
Tetrose	Tetrulose	1	—	—
Pentose	Pentulose	2	Pentulofuranose	—
Hexose	Hexulose	3	Hexulofuranose	Hexulopyranose
Heptose	Heptulose	4	Heptulofuranose	Heptulopyranose
Octose	Octulose	5	Octulofuranose	Octulopyranose
Nonose	Nonulose	6	Nonulofuranose	Nonulopyranose
Decose	Deculose	7	Deculofuranose	Deculopyranose

[a] Although an achiral aldobiose (glycolaldehyde) exists, it is not considered to be a saccharide because, by definition, a saccharide must contain at least one asymmetric carbon atom.

[b] Although an achiral triulose (1,3-dihydroxyacetone) exists, it is not considered to be a saccharide because it lacks an asymmetric carbon atom.

mers of monosaccharides linked to a nonsaccharide, such as a lipid or a peptide. One can further classify simple oligosaccharides, according to degree of polymerization, into disaccharides, trisaccharides, tetrasaccharides, etc., and according to whether or not the oligomer chain has at one end a hemiacetal function (a latent aldehyde or keto group). Such terminal groups are readily converted into carboxylic groups by mild oxidants and, accordingly, oligosaccharides possessing these groups are referred to as *reducing,* in contradistinction to those which resist such oxidation and are designated *nonreducing*. Thus, whereas all monosaccharides are reducing, there are reducing and nonreducing disaccharides, trisaccharides, etc. Because of their sweet taste, monosaccharides and lower oligosaccharides have been called sugars. It should be noted, however, that sweetness decreases with increasing DP of oligosaccharides, and beyond a DP of 4 the oligomer is tasteless. Table II shows the classification of oligosaccharides and gives the names of the most common ones.

C. Polysaccharides

Polysaccharides and oligosaccharides are polymeric in nature and are structurally similar (both are polyacetals having oxygen bridges linking the monosaccharide monomers), but they may differ markedly in degree

TABLE II
Oligosaccharides[a]

Reducing		Nonreducing	
Homo-oligosaccharides	Hetero-oligosaccharides	Homo-oligosaccharides	Hetero-oligosaccharides
	Simple oligosaccharides		
I. Disaccharides			
Maltose	Lactose	Trehalose	Sucrose
4-α-D-Glc*p*-D-Glc	4-β-D-Gal*p*-D-Glc	α-D-Glc*p*-α-D-Glc*p*	β-D-Fru*f*-α-D-Glc*p*
Cellobiose	Lactulose	Isotrehalose	Isosucrose
4-β-D-Glc*p*-D-Glc	4-β-D-Gal*p*-D-Fru	β-D-Glc*p*-β-D-Glc*p*	α-D-Fru*f*-β-D-Glc*p*
Isomaltose	Melibose		
6-α-D-Glc*p*-D-Glc	6-α-D-Gal*p*-D-Glc		
Gentiobiose	Turanose		
6-β-D-Glc*p*-D-Glc	3-α-D-Glc*p*-D-Fru		
II. Trisaccharides			
Maltotriose	Manninotriose		Raffinose
4-α-D-Glc*p*-maltose	6-α-D-Gal*p*-melibiose		6-α-D-Gal*p*-sucrose
III. Tetrasaccharides			
Maltotetraose			Stachyose
4-α-D-Glc*p*-maltotriose			6-α-D-Gal*p*-raffinose
IV. Pentasaccharides			
V. Hexasaccharides			
VI. Heptasaccharides			
VII. Octasaccharides			
VIII. Nonasaccharides			
IX. Decasaccharides			
	Conjugate oligosaccharides		
			Glycolipids

[a] Glc, Glucose; Gal, galactose; Fru, fructose; *f*, furanose; *p*, pyranose.

of polymerization; the polysaccharides may reach a DP of 10^5, whereas, by definition, the maximum DP for oligosaccharides is 10. Although, by convention, compounds having a degree of polymerization of 11 or more are designated polysaccharides, the differences between the properties of lower polysaccharides with a DP of 11 and those of higher (DP = 10) oligosaccharides can hardly be detected. However, most polysaccharides have a much higher degree of polymerization than oligosaccharides, which renders quite significant the sum of the gradual changes that occur in their physical properties with increasing DP. For example, because the solubility decreases and the viscosity increases with a rise in DP, some higher polysaccharides, such as cellulose, are completely insoluble in water (all oligosaccharides are soluble), while the increase in viscosity may cause the solutions of other polysaccharides to set and gel.

There are several ways in which to classify polysaccharides. A common one is to group them according to their sources, that is, into plant and animal polysaccharides, and then subdivide the former into skeletal polysaccharides (cellulose, etc.), reserve polysaccharides (starch, etc.), gums and mucilages, algal polysaccharides, bacterial polysaccharides, and so forth. The disadvantage of this classification is that it tells us very little about the chemistry of these polymers.

The classification used in chemistry texts distinguishes between (a) simple (true) polysaccharides, which afford on depolymerization only mono- and oligosaccharides or their derivatives (esters or ethers), and (b) conjugate polymers made up of a polysaccharide linked to another polymer, such as a peptide or a protein (to form a glycopeptide or glycoprotein). Polysaccharides are, in turn, grouped into two major classes: (i) homopolysaccharides, which are simple polymers having as a repeating unit (monomer) one type of monosaccharide, and (ii) heteropolysaccharides, which are made up of more than one type of monosaccharide. Because the shape of polymers significantly influences their physical properties, each of these types of polymer is further divided into linear and branched polysaccharides. Table III shows the classes to which some of the common polysaccharides belong.

D. DNA, RNA, Nucleotides, and Nucleosides

Unlike oligo- and polysaccharides, which are polyacetals linked by oxygen bridges, DNA and RNA are polyesters linked by phosphate bridges. DNA is the largest known polymer; its DP exceeds 10^{12} in human genes and decreases as the ladder of evolution is descended. This giant molecule plays a key role in replication and in transcription. The latter is achieved by doubling one of the DNA strands with a smaller polymer,

TABLE III
Polysaccharides

Homopolysaccharides		Heteropolysaccharides	
Linear	Branched	Linear	Branched
	Simple polysaccharides		
Amylose (α-D-glucan)	Amylopectin	Mannans	Gums
Cellulose (β-D-glucan)	Glycogen	Xylans	Mucilages
Chitin (D-glucosaminan)			Pectins
			Algin
			Agar
			Bacterial polysaccharides
	Conjugate polysaccharides		
			Peptidoglycans
			Glycoproteins
			Lectins

mRNA, which in turn binds with a string of oligomers, tRNA, to form the peptide chain. The monomers of DNA and RNA are made up of phosphorylated 2-deoxy-D-*erythro*-pentofuranosyl- and D-ribofuranosyl-purine and -pyrimidine bases, designated nucleotides. The latter can undergo hydrolysis of their phosphoric ester groups to afford simpler monomers, the nucleosides. Thus, it is apparent that this group may also be divided, according to the degree of polymerization, into monomers (nucleosides and nucleotides), oligomers (tRNA), and polymers (DNA and mRNA).

E. Other Saccharide Derivatives

Grouped under this broad heading are a number of important carbohydrate derivatives which are best studied together, rather than being artificially divided to conform to a certain classification. For example, carbohydrate-containing antibiotics constitute a group of complex organic molecules of therapeutic importance. Some of their members are monosaccharide derivatives and, if grouped under this class, would be separated from closely related oligosaccharide derivatives.

I

Monosaccharides

2

Structure, Configuration, and Conformation of Monosaccharides

Monosaccharides (monomeric sugars) exist as chiral polyhydroxyalkanals, called aldoses, or chiral polyhydroxyalkanones, called ketoses. These are further classified (see Table I), according to the number of carbon atoms in their chains, into trioses, tetroses, pentoses, hexoses, etc. and, according to the type of ring they form, into furanoses and pyranoses (five- and six-membered rings).

D-Glucose is the most abundant monosaccharide and therefore the most extensively studied. In this chapter it will be used as an example to illustrate how the structure, configuration, and conformation of monosaccharides can be determined. Although reference may be made to how this was originally done, the emphasis will be on the methods used today to elucidate the structures of monosaccharides and their derivatives.

I. STRUCTURE OF MONOSACCHARIDES

The first point to be determined in elucidating the structure of an organic compound is the molecular formula, which is achieved by the combined use of elemental (combustion) analysis and determination of molecular weight. The most accurate method available for molecular weight

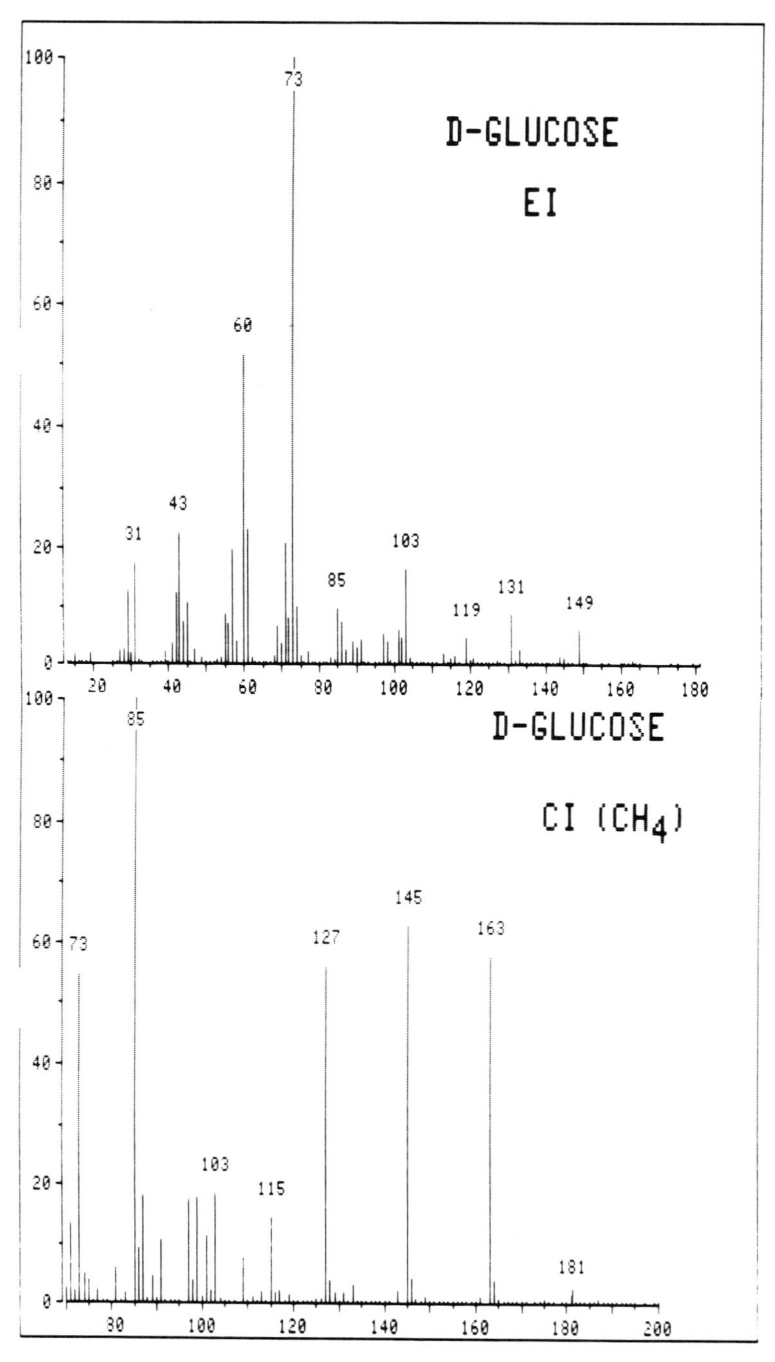

Fig. 1. Mass spectra of D-glucose, using electron impact (top) and chemical ionization (bottom).

determination is mass spectrometry (MS). For carbohydrates, chemical ionization mass spectrometry (CI-MS) is the preferred method (see page 81). The advantage of this type of MS over the simpler, electron impact mass spectrometry (EI-MS), is that it requires less energy to ionize the molecule, which increases the chances of observing the molecular peak (M^+) in sensitive molecules. The higher energy used in EI-MS causes considerable fragmentation of the carbon chain and produces a mass spectrum from which the molecular peak may be absent. The EI and CI mass spectra of D-glucose are depicted in Fig. 1. It shows an (M + 1) peak (*m*/*z* 181), arising from the transfer of a proton from the methane gas used for ionization, indicating that the molecular weight of D-glucose is 180. On the other hand, the highest mass observed in the EI-MS depicted is at 149, which corresponds to loss of the terminal CH_2OH.

A molecular weight of 180 for D-glucose, in combination with the elemental analysis results, which agree with an empirical formula of CH_2O (molecular weight = 30), would suggest that the molecular formula is $C_6H_{12}O_6$.

The next step in elucidating the structure of a monosaccharide is to determine whether the carbon atoms in the chain are arranged in a linear or branched manner. This can be done by converting a new saccharide into a derivative of established structure. At the turn of the century, glucose was converted into 1-iodohexane by treatment with HI, which demonstrated that the six carbon atoms of this sugar were arranged in a linear fashion.

To determine the number of hydroxyl groups in a monosaccharide molecule, the latter can be converted into the acetate and this subjected to an acetyl number analysis. This involves titration of the alkali remaining after saponification with a known excess of the base. A simpler method is actually to count the number of (acetyl) methyl protons in the ^1H nuclear magnetic resonance spectrum. The ^1H-NMR spectrum of peracetylated D-glucose (penta-O-acetyl-β-D-glucopyranose) in deuteriochloroform (see Fig. 2) clearly shows five such methyl proton peaks, suggesting the presence of five hydroxyl groups in the parent D-glucose.

The presence of an aldehyde group in D-glucose was established early, from the fact that D-glucose readily reacts with carbonyl group reagents, giving hydrazones with hydrazines and an oxime with hydroxylamine. Furthermore, because D-glucose is readily oxidized with mild oxidants to an acid (D-gluconic acid) that possesses the same number (six) of carbon atoms, it was concluded that it possesses a terminal carbonyl group, i.e., an aldehyde group. By using the data available so far, it is possible to deduce for the acyclic form of D-glucose a structural formula (**1**) that agrees with the valence requirements of the various elements therein:

$$
\begin{array}{c}
\mathrm{HC{=}O} \\
| \\
\mathrm{(CHOH)_4} \\
| \\
\mathrm{CH_2OH}
\end{array}
$$

1

It should be noted at this point that a carbonyl band is barely visible in the infrared spectrum of D-glucose and that an aldehyde proton can hardly be detected in its NMR spectrum, because this sugar, like most monosaccharides, exists in solution almost exclusively in cyclic forms. These forms are produced by nucleophilic attack of the oxygen atom of one of the hydroxyl groups on the carbonyl group, which is thus converted into a hemiacetal. Detailed discussion of the cyclic forms of D-glucose will be deferred until the configuration of its acyclic form has been presented.

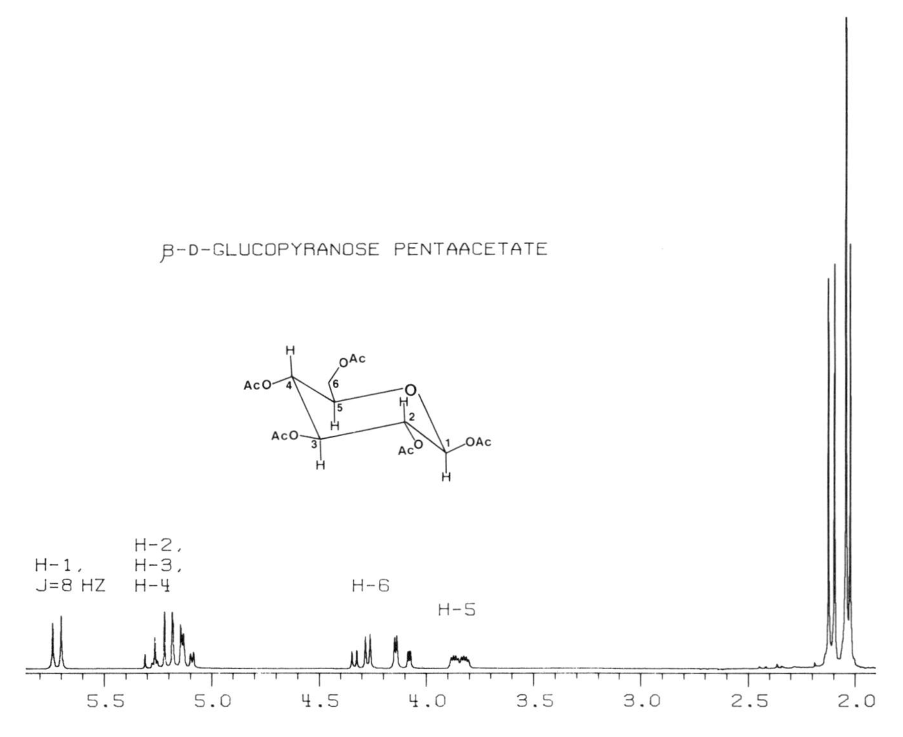

Fig. 2. ^{1}H-NMR spectrum of penta-O-acetyl-β-D-glucopyranose in deuteriochloroform.

II. CONFIGURATION OF ACYCLIC MONOSACCHARIDES

A. The Fischer Projection

Before discussing the configuration of monosaccharides, it will be useful to describe in some detail one of the conventional methods used to represent three-dimensional molecules two-dimensionally. The system commonly used for *linear* monosaccharides is the Fischer projection formula, which affords an unambiguous way to depict such molecules, provided they are oriented as follows: (a) the carbon chain is drawn vertically, with the carbonyl group at (or nearest to) the top and the last carbon atom in the chain, i.e., the one farthest from the carbonyl group, at the bottom; (b) each carbon atom is rotated around its vertical axis until all of the vertical (C–C) bonds in the chain lie below an imaginary curved plane (which will later be represented by the surface of the paper) and all of the horizontal bonds (parallel to the *x*-axis) lie above this plane and are directed toward the observer. The curved plane is then flattened, and the projection of the molecule is drawn as viewed.

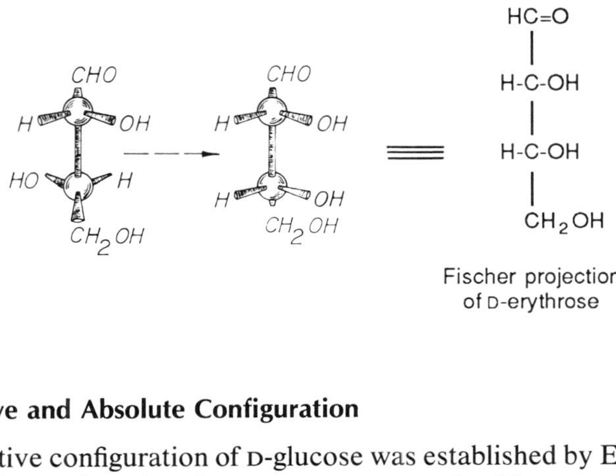

Fischer projection of D-erythrose

B. Relative and Absolute Configuration

The relative configuration of D-glucose was established by Emil Fischer in 1891 and constituted at the time a monumental achievement, for which he earned a Nobel prize. Today, the determination of the absolute configuration of a monosaccharide offers no difficulty, because a large number of related compounds of known configuration are available. The unknown is simply converted into a compound of known configuration, using a reaction sequence that does not affect the configuration at the chiral center(s).

From structure **1** it may be seen that an aldohexose such as D-glucose possesses four chiral carbon atoms; that is, it exists as $2^4 = 16$ stereoisomers. These may be grouped into eight pairs of enantiomers having the following (*R*) and (*S*) configurations:

RRRR	*SRRR*	*RSRR*	*SSRR*	*RRSR*	*SRSR*	*RSSR*	*SSSR*
and	and	and	and	and	and	and	and
SSSS	*RSSS*	*SRSS*	*RRSS*	*SSRS*	*RSRS*	*SRRS*	*RRRS*

To determine which of the above combinations corresponds to that of D-glucose, it is necessary to have a reference compound of known absolute configuration that can be prepared from, or converted into, D-glucose. Because the first determination of the absolute configuration of an organic compound had to await the advent of X-ray crystallographic techniques developed in the middle of the twentieth century, such a reference compound did not exist in Fischer's time. This is why he was able only to propose a relative configuration for the monosaccharides known in his time. To do this, he used as the reference compound the dextrorotatory form of glyceraldehyde, now designated D-(+)-glyceraldehyde (**2**), which can be obtained from D-glucose by repeated degradations and can, in turn, yield D-glucose by repeated ascending syntheses. He arbitrarily proposed that the OH attached to C-2 in this compound is to the right when represented by a Fischer projection formula [it has the (*R*) configuration].

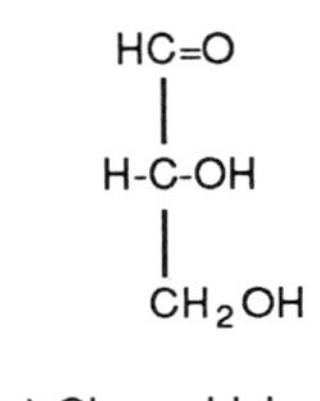

D-(+)-Glyceraldehyde

2

It was fortunate that the arbitrary assignment of structure **2** to the enantiomer of glyceraldehyde related to D-glucose was later found by X-ray crystallography to be the correct one. Otherwise, all of the configurations assigned to carbohydrates during the 60 years that followed Fischer's assignment would have had to be reversed.

Once the absolute configuration at C-2 of the D-(+) isomer of glyceraldehyde had been established, it could be related to the configuration of the corresponding center in any aldose that is obtainable from this isomer by ascending the series or that yields this isomer by repeated descending degradations.

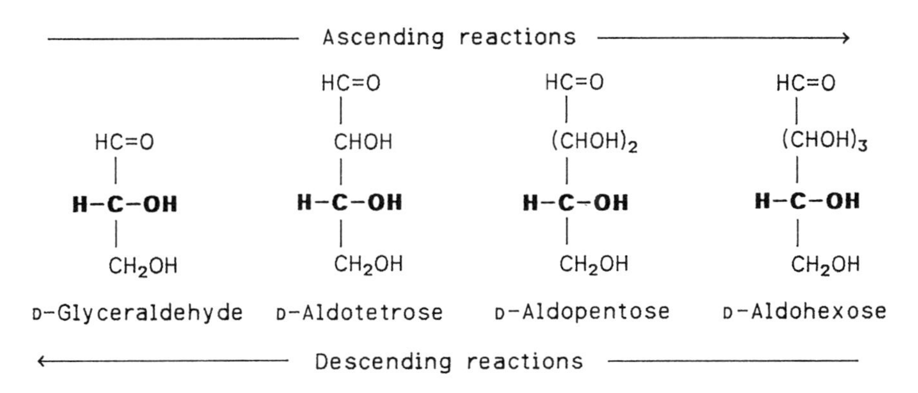

C. Ascending and Descending the Series

To illustrate how ascending and descending reactions are performed, an example of each will be briefly outlined. A widely used method of ascending the series, developed by H. Kiliani, is the cyanohydrin synthesis, carried out by initiating a nucleophilic attack on the carbonyl group with HCN, hydrolyzing the nitrile formed to a lactone, and then reducing the latter to the two higher aldoses.

HC=O / H–C–OH / CH_2OH (D-Triose) ⟶ CN / CHOH / H–C–OH / CH_2OH (Nitrile) ⟶ C=O / CHOH / H–C–OH / CH_2O (1,4-Lactone) ⟶ HC=O / CHOH / H–C–OH / CH_2OH (D-Tetrose)

A reaction used to descend the series is the Ruff degradation, which involves oxidation of an aldose first with bromine water, to the aldonic acid, and then further with peroxide, to afford an α-keto carboxylic acid. The latter then undergoes spontaneous decarboxylation, affording the lower aldose in the series.

D. D and L Families

Two consecutive cyanohydrin syntheses are needed to convert D-(+)-glyceraldehyde (**2**) into D-arabinose (and other pentoses of the D family),

L-Aldopentose	→	Aldonic acid	→	α-Keto acid	→	L-Tetrose
HC=O		COOH		COOH		
H-C-OH		H-C-OH		C=O		HC=O
HO-C-H		HO-C-H		HO-C-H		HO-C-H
HO-C-H		HO-C-H		HO-C-H		HO-C-H
CH_2OH		CH_2OH		CH_2OH		CH_2OH

and an additional synthesis is needed to afford D-glucose, D-mannose, and the other aldoses of the D family (see Fig. 3). Conversely, a Ruff degradation is needed to convert D-glucose or D-mannose into D-arabinose, and two further degradations are required to convert D-arabinose into D-(+)-glyceraldehyde. From this, it may be concluded that C-2 of glyceraldehyde corresponds to C-4 of arabinose and to C-5 of the two hexoses mentioned. Furthermore, it may be concluded that all of the aforementioned chiral centers must have the same configuration.

D-Aldotriose	→ →	D-Aldopentose	→	D-Aldohexose
				HC=O
		HC=O		-C-
		-C-		-C-
HC=O		-C-		-C-
H-C-OH		H-C-OH		H-C-OH
CH_2OH		CH_2OH		CH_2OH

It should be stated at this point that, because C-2 in glyceraldehyde corresponds to the chiral center farthest from the carbonyl group in any aldose or ketose, it is possible to group monosaccharides into two families: one related to D-(+)-glyceraldehyde and designated by the prefix D, and one related to L-(−)-glyceraldehyde and designated by the prefix L. Monosaccharides of the D family all have the (*R*) configuration at the chiral center farthest from the carbonyl (the OH attached to this chiral carbon is to the right in a Fischer projection). Conversely, monosaccharides of the L family have the (*S*) configuration at this center (the OH group is to the left).

E. Configuration of the Acyclic Form of D-Glucose

Fischer used an ingenious method to determine whether, in an aldose molecule, the chiral centers that are equidistant from the center (for example, C-2 and C-4 in a pentose) have the same configuration. He determined whether conversion of the two groups (CHO and CH_2OH) situated at the "top" and the "bottom" of an aldose molecule into the same type

(a)

D-Glyceraldehyde
HC=O
H-C-OH
CH_2OH

D-Erythrose	D-Threose
HC=O	HC=O
H-C-OH	HO-C-H
H-C-OH	**H-C-OH**
CH_2OH	CH_2OH

D-Ribose	D-Arabinose	D-Xylose	D-Lyxose
HC=O	HC=O	HC=O	HC=O
H-C-OH	HO-C-H	H-C-OH	HO-C-H
H-C-OH	H-C-OH	HO-C-H	HO-C-H
H-C-OH	**H-C-OH**	**H-C-OH**	**H-C-OH**
CH_2OH	CH_2OH	CH_2OH	CH_2OH

D-Allose	D-Altrose	D-Glucose	D-Mannose	D-Gulose	D-Idose	D-Galactose	D-Talose
HC=O	HC=O	HC=O	HC=O	HC=O	HC=O	HC=O	HC=O
H-C-OH	HO-C-H	H-C-OH	HO-C-H	H-C-OH	HO-C-H	H-C-OH	HO-C-H
H-C-OH	H-C-OH	HO-C-H	HO-C-H	H-C-OH	H-C-OH	HO-C-H	HO-C-H
H-C-OH	H-C-OH	H-C-OH	H-C-OH	HO-C-H	HO-C-H	HO-C-H	HO-C-H
H-C-OH	**H-C-OH**	**H-C-OH**	**H-C-OH**	**H-C-OH**	**H-C-OH**	**H-C-OH**	**H-C-OH**
CH_2OH	CH_2OH	CH_2OH	CH_2OH	CH_2OH	CH_2OH	CH_2OH	CH_2OH

Fig. 3. (a), Aldoses and (b), ketoses of the D family.

(b)

```
CH2OH
  |
 C=O
  |
H-C-OH
  |
CH2OH
```

D-*glycero*-Tetrulose

```
CH2OH
  |
 C=O
  |
H-C-OH
  |
H-C-OH
  |
CH2OH
```

D-*erythro*-Pentulose

```
 CH2OH
   |
  C=O
   |
HO-C-H
   |
 H-C-OH
   |
 CH2OH
```

D-*threo*-Pentulose

```
CH2OH
  |
 C=O
  |
H-C-OH
  |
H-C-OH
  |
H-C-OH
  |
CH2OH
```

D-*ribo*-Hexulose
D-Psicose

```
 CH2OH
   |
  C=O
   |
HO-C-H
   |
 H-C-OH
   |
 H-C-OH
   |
 CH2OH
```

D-*arabino*-Hexulose
D-Fructose

```
 CH2OH
   |
  C=O
   |
 H-C-OH
   |
HO-C-H
   |
 H-C-OH
   |
 CH2OH
```

D-*xylo*-Hexulose
D-Sorbose

```
 CH2OH
   |
  C=O
   |
HO-C-H
   |
HO-C-H
   |
 H-C-OH
   |
 CH2OH
```

D-*lyxo*-Hexulose
D-Tagatose

Fig. 3. (*Continued.*)

of group, for example, a carboxylic group, rendered the product achiral (afforded a *meso* compound). If so, he could conclude that a plane of symmetry was created during the conversion (oxidation) and that the configurations of the chiral centers situated at equal distances from this plane of symmetry were identical. Because the dicarboxylic acids he needed had already been prepared by oxidation of aldoses with nitric acid, Fischer did not have to carry out the oxidation himself; he looked up the literature to find out whether the resulting acids were optically active or not. An alternative method that could be used is reduction of the aldehyde group to a primary hydroxyl group, which would afford an alditol (a polyol) having identical groups (CH_2OH) at both ends of the molecule. As all of the aldoses are chiral (and therefore do not have a plane of symmetry), the formation of an optically inactive alditol, like the formation of an optically inactive polyhydroxydicarboxylic acid, would indicate that the product had acquired a plane of symmetry.

Using the reasoning just outlined, Emil Fischer was able in four steps to determine the configuration of a pentose, D-arabinose, and in five steps that of three hexoses, D-glucose, D-mannose, and L-gulose. The following is a modern rendition of Fischer's famous elucidation of the configuration of the acyclic forms of these sugars, adapted to conform with present knowledge.

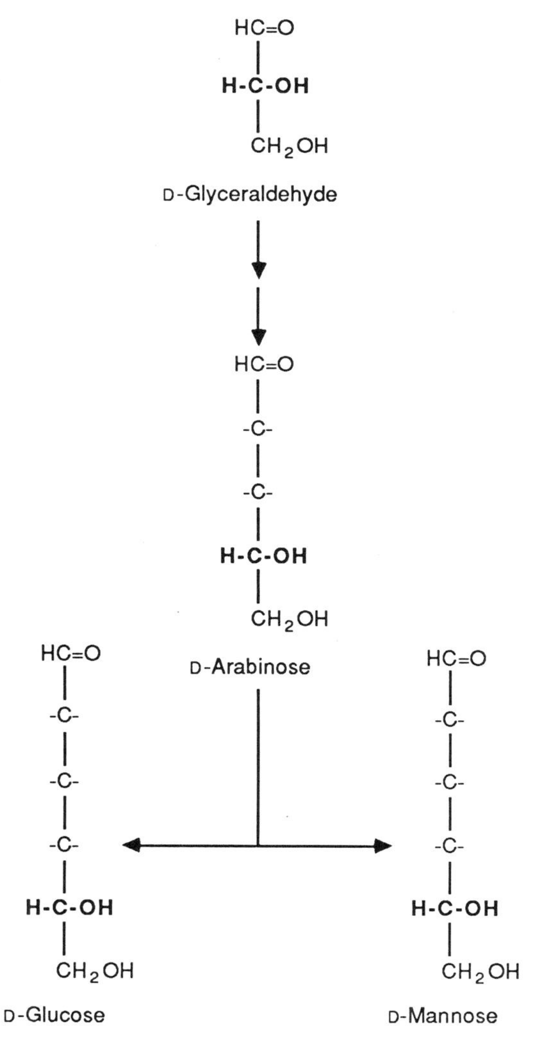

1. Because X-ray crystallography has confirmed that the configuration of C-2 in D-(+)-glyceraldehyde is (*R*), as arbitrarily assigned by Fischer, and because C-2 in the latter compound is equivalent to C-4 in such D-pentoses as D-arabinose and to C-5 in such D-hexoses as D-glucose and D-mannose, it follows that the configuration of C-4 and C-5 in D-arabinose and in the two hexoses (D-glucose and D-mannose) is likewise (*R*) (as they all have the OH group to the right in a Fischer projection).

2. D-Arabinose yields a chiral (optically active) dicarboxylic acid, D-arabinaric acid, on oxidation with nitric acid and a chiral pentitol (D-arabinitol) on reduction. Accordingly, none of these reaction products is a *meso* compound (the latter, by definition, necessitates a plane of symmetry, which would have negated the chirality). To agree with this fact, it must be concluded that C-2 in D-arabinose has the (*S*) configuration, as otherwise the dicarboxylic acid and the pentitol would possess a plane of symmetry and be achiral. Because C-2 in D-arabinose corresponds to C-3 in the two D-hexoses, both of these chiral carbon atoms must likewise have (*S*) configurations, and their OH groups must be to the left.

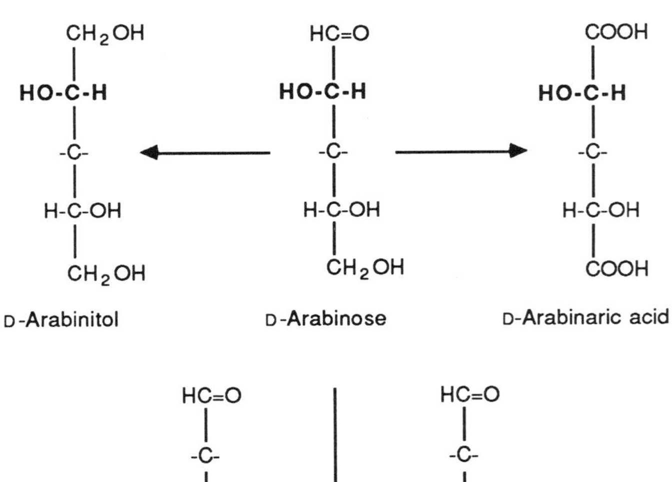

HC=O
-C-
HO-C-H
-C-
H-C-OH
CH2OH

HC=O
-C-
HO-C-H
-C-
H-C-OH
CH2OH

D-Glucose D-Mannose

It may be noted that if C-2 in D-arabinose had the (*R*) configuration (with its OH group to the right), both the dicarboxylic acid and the alditol would have been *meso* compounds (optically inactive).

```
 CH2OH              HC=O              COOH
   |                  |                 |
 H-C-OH            H-C-OH            H-C-OH
   |                  |                 |
  -C-   <--------    -C-   -------->   -C-
   |                  |                 |
 H-C-OH            H-C-OH            H-C-OH
   |                  |                 |
 CH2OH              CH2OH             COOH
```

3. Because D-arabinose affords a mixture of D-glucose and D-mannose in the cyanohydrin synthesis (nucleophilic attack by HCN on the carbonyl group, followed by hydrolysis of the resulting nitriles and reduction of the lactone), it may be concluded that the two hexoses are 2-epimers; that is, they differ only in the configuration at C-2, one being (*R*) and the other (*S*). Without identifying which is which, the 2-hydroxyl group can be drawn to the right in one hexose and to the left in the other.

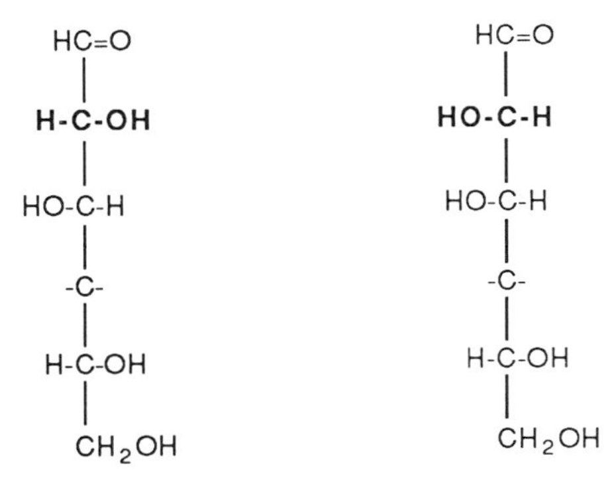

D-Glucose (or D-mannose) D-Glucose (or D-mannose)

4. Both D-glucose and D-mannose afford chiral dicarboxylic acids (D-glucaric acid and D-mannaric acid), as well as chiral hexitols (D-glucitol and D-mannitol). It must follow that C-4 in the two hexoses has the (*R*) configuration (the OH group at C-4 is to the right), otherwise one of the dicarboxylic acids and one of the hexitols would have been achiral (*meso*). Because C-4 in a hexose corresponds to C-3 in a related pentose, it must be concluded that, in D-arabinose, C-3 is likewise of the (*R*) configuration (the OH at C-3 in D-arabinose is to the right). At this stage, the configuration of the acyclic form of D-arabinose has been completely

established, and what remains is to determine which of the two hexose structures corresponds to D-glucose and which to D-mannose.

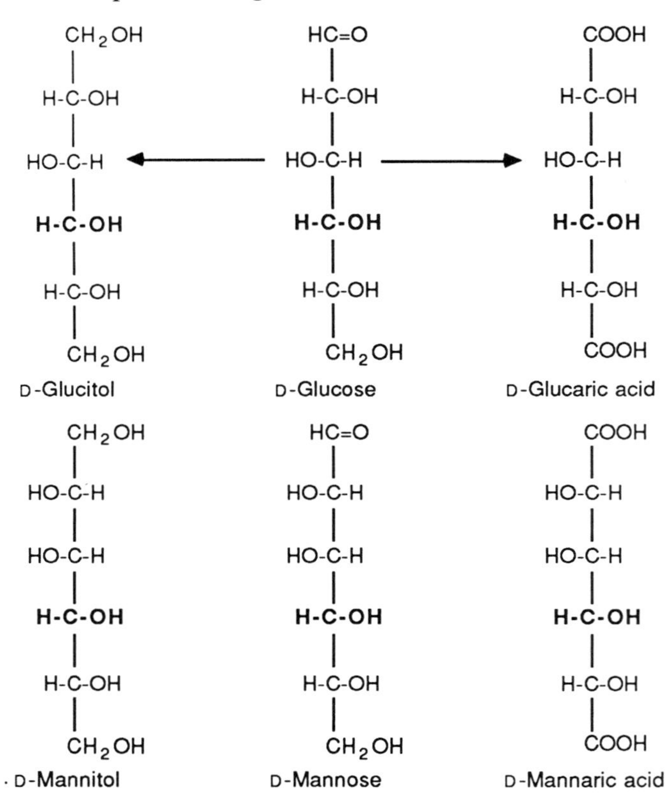

None of the foregoing alditols and dicarboxylic acids is a *meso* compound. If, however, C-4 were (*S*), *meso* products would result.

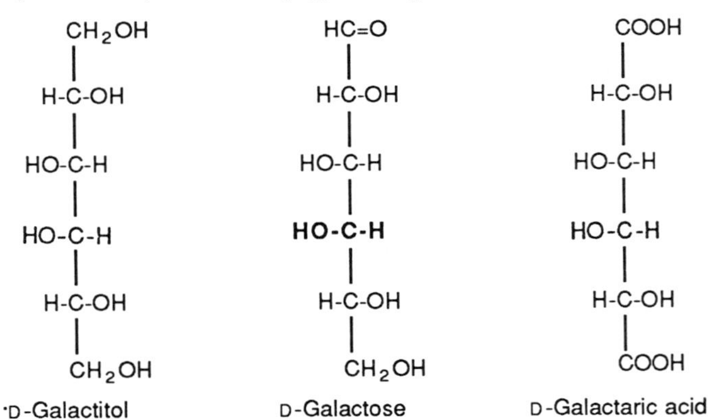

This now establishes the structure of D-arabinose as follows.

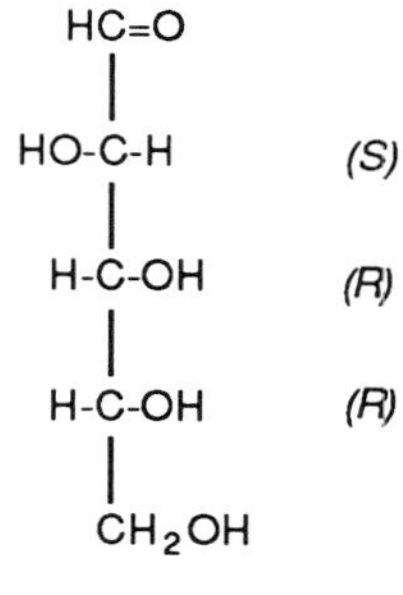

D-Arabinose

5. If a polyhdroxydicarboxylic acid or a polyol possesses an axis of symmetry, it can be obtained from only one aldose, whereas if it lacks such an axis, it can be obtained from two different aldoses. In the first case, each rotation of 180° around the axis of symmetry will produce the same arrangement of groups. In the second case, with no axis of symmetry, the same acid or polyol may be produced from two aldoses: an aldose having the aldehyde group at the "top" and the primary hydroxyl group at the "bottom," and another aldose having the aldehyde group at the bottom and the primary hydroxyl group at the top, which is, of course, inconsistent with the Fischer projection rules, and the molecule must be rotated 180° in order to represent it correctly.

As it is known that the dicarboxylic acids, and the hexitols, obtained from D-glucose and L-gulose are identical (D-glucaric acid is identical to L-gularic acid, and D-glucitol is identical to L-gulitol), whereas the dicarboxylic acid and the hexitol obtained from D-mannose (D-mannaric acid and D-mannitol) are not obtained from any other aldose, it must be concluded that neither D-glucaric acid nor D-glucitol has an axis of symmetry and that D-mannaric acid and D-mannitol must each possess an axis of symmetry. It follows from the first conclusion, namely that the dicarboxylic acid and the hexitol obtained from D-glucose and L-gulose do not have an axis of symmetry, that C-2 in both the acid and the hexitol must have the (*R*) configuration (the OH at this center is to the right). Of course, the same (*R*) configuration must also be assigned to C-2 of D-glucose and C-4 of L-gulose (the latter atom corresponds to C-2 of D-glucose). Using the same reasoning, it must be concluded that both D-mannaric acid and D-mannitol possess an axis of symmetry and must therefore have the (*S*) configuration at C-2 (the OH is to the left). The same configuration (*S*) must now be assigned to C-2 of D-mannose, and thus this concludes the elucidation of the total configuration of the acyclic forms of D-glucose and D-mannose, as well as of L-gulose.

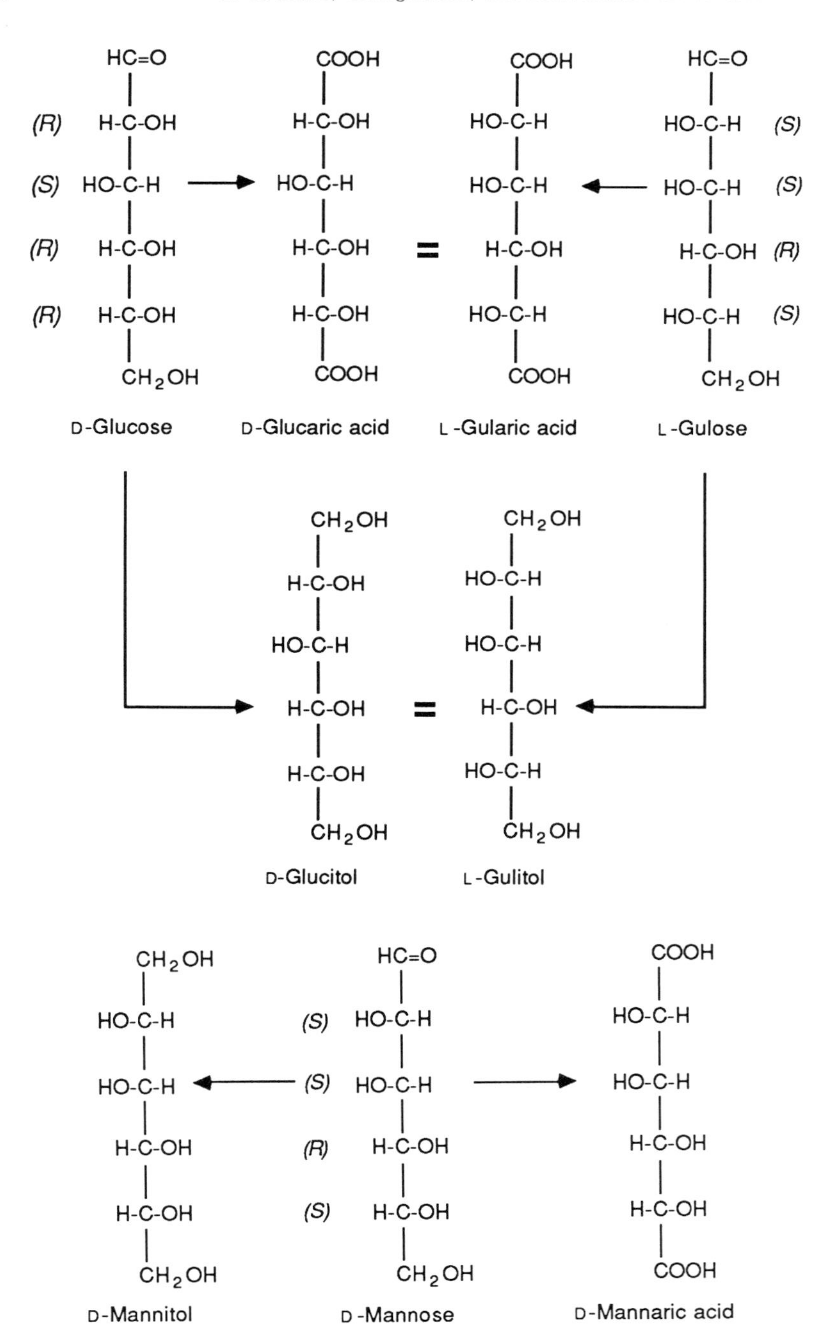
HC=O
(R) H-C-OH
(S) HO-C-H
(R) H-C-OH
(R) H-C-OH
CH2OH
D-Glucose
COOH
H-C-OH
HO-C-H
H-C-OH
H-C-OH
COOH
D-Glucaric acid
=
COOH
HO-C-H
HO-C-H
H-C-OH
HO-C-H
COOH
L-Gularic acid
HC=O
HO-C-H (S)
HO-C-H (S)
H-C-OH (R)
HO-C-H (S)
CH2OH
L-Gulose
CH2OH
H-C-OH
HO-C-H
H-C-OH
H-C-OH
CH2OH
D-Glucitol
=
CH2OH
HO-C-H
HO-C-H
H-C-OH
HO-C-H
CH2OH
L-Gulitol
CH2OH
HO-C-H
HO-C-H
H-C-OH
H-C-OH
CH2OH
D-Mannitol
HC=O
(S) HO-C-H
(S) HO-C-H
(R) H-C-OH
(S) H-C-OH
CH2OH
D-Mannose
COOH
HO-C-H
HO-C-H
H-C-OH
H-C-OH
COOH
D-Mannaric acid

Soon after the configurations of the foregoing aldoses had been determined, the configurations of other aldoses and ketoses were established rapidly by using correlations with one or more of the known configurations.

III. CYCLIC STRUCTURES, RING SIZE, AND ANOMERIC CONFIGURATION

Soon after the configuration of the acylic form of glucose and the other aldoses was established, it became apparent that these structures could not be the only ones in existence, nor could they constitute the major components in an equilibrium mixture. Thus, two forms of D-glucose were isolated, instead of only one (as would be expected from the acyclic structure). These forms were designated α and β (a third one was also isolated, but it proved to be an equilibrium mixture). Each of the two forms has, in solution, a characteristic optical rotation that changes with time until it reaches a constant value (that of the equilibrium mixture). The change in optical rotation with time is called *mutarotation,* and it is indicative of molecular rearrangements occurring in solution.

	α-D-glucose	$\rightleftharpoons$ equilibrium mixture $\rightleftharpoons$	β-D-glucose
$[\alpha]_D$	+ 112.2°	+ 52.7°	+ 18.7°

Evidence that the acyclic structure of D-glucose is not the major component of the equilibrium mixture was mentioned briefly on page 14, where it was stated that IR and NMR spectroscopy failed to detect an aldehyde group in D-glucose. Thus, the IR spectrum of β-D-glucopyranose shown in Fig. 4 does not exhibit a carbonyl band at 1700 cm^{-1}, characteristic of such groups. Furthermore, the ^{1}H-NMR spectrum of α-D-glucose, shown in Fig. 5, lacks and aldehydic proton at δ 10, also characteristic of such a group.

It is now known that D-glucose and the other aldoses that have the necessary number of carbon atoms exist almost exclusively in the form of five- or six-membered cyclic hemiacetals. These forms are the outcome of an intramolecular nucleophilic attack by the hydroxyl oxygen atom attached to C-4 or C-5 on the carbonyl group. The five-membered ring produced in the first case is related to tetrahydrofuran and is therefore designated *furanose*. The six-membered ring produced in the second case is related to tetrahydropyran and is called *pyranose*. Because cyclization converts an achiral aldehyde carbon atom into a chiral hemiacetal carbon atom, two isomers, termed anomers, are produced; these are designated α and β. Accordingly, a solution of an aldose at equilibrium contains a

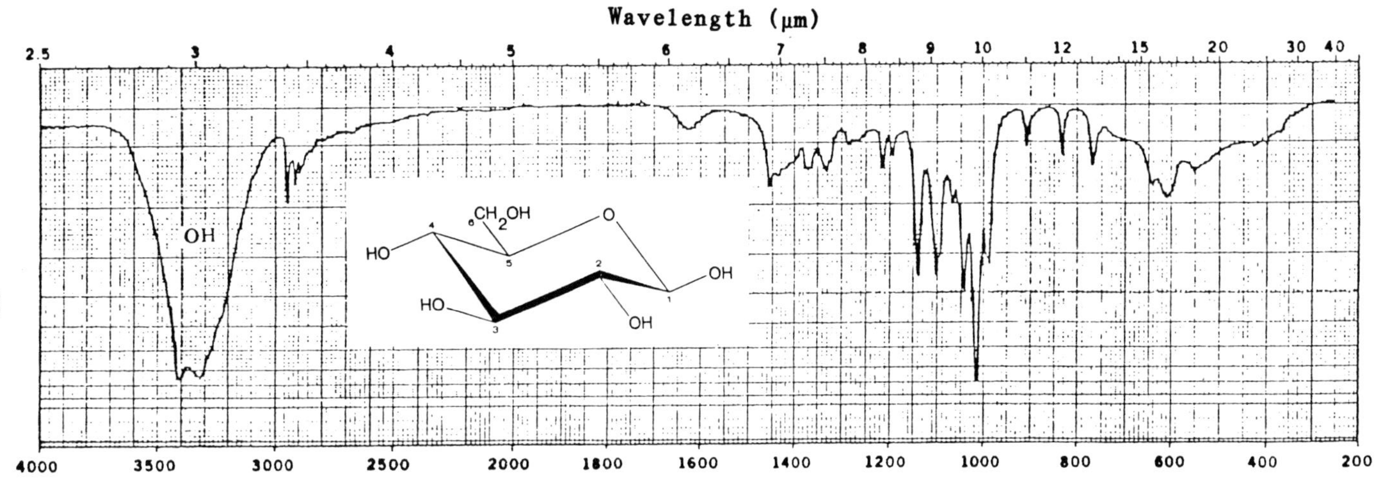

Fig. 4. Infrared spectrum of crystalline β-D-glucopyranose (KBr).

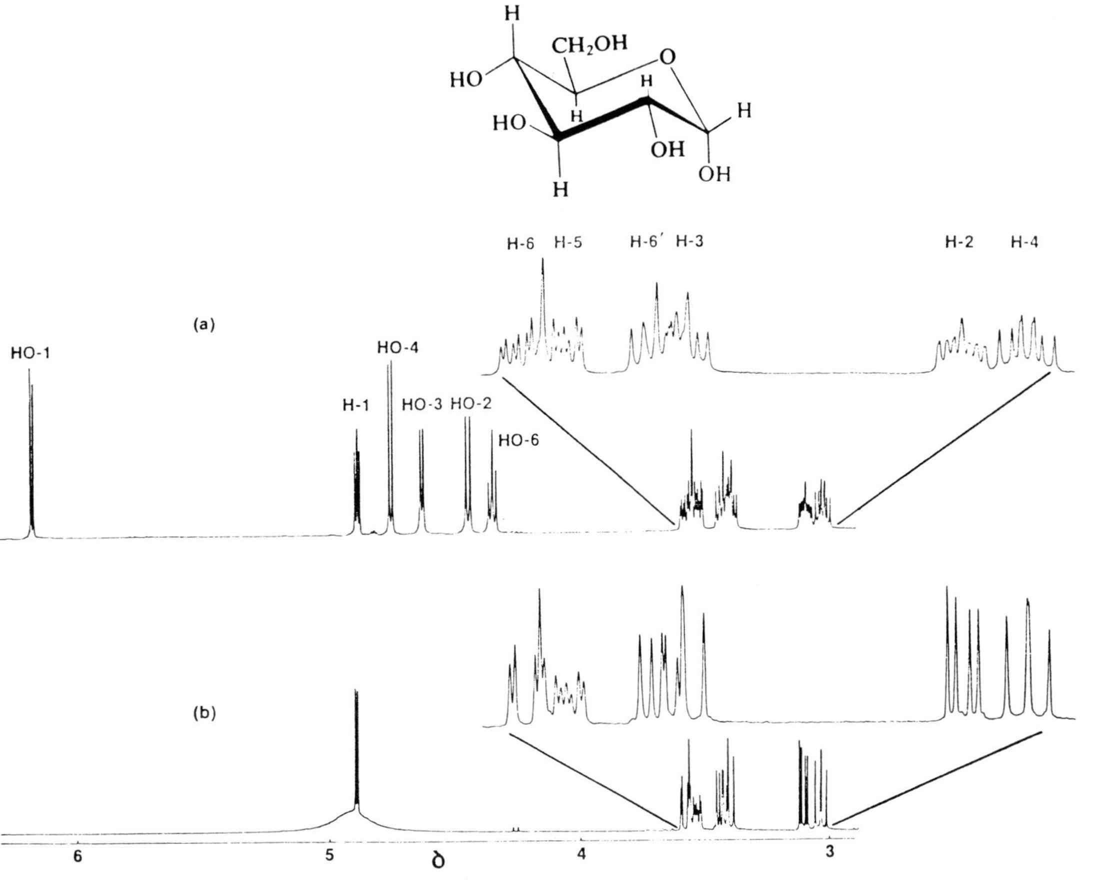

Fig. 5. ^{1}H-NMR spectra of a solution of α-D-glucose in Me_2SO-d_6 at 400 MHz. (a) OH protons coupled; (b) OH protons decoupled.

mixture of at least two (α and β) furanoses and two (α and β) pyranoses, as well as possibly traces of the acyclic form and its hydrate. In addition, α and β septanose forms may exist, but they are not detectable with aldoses.

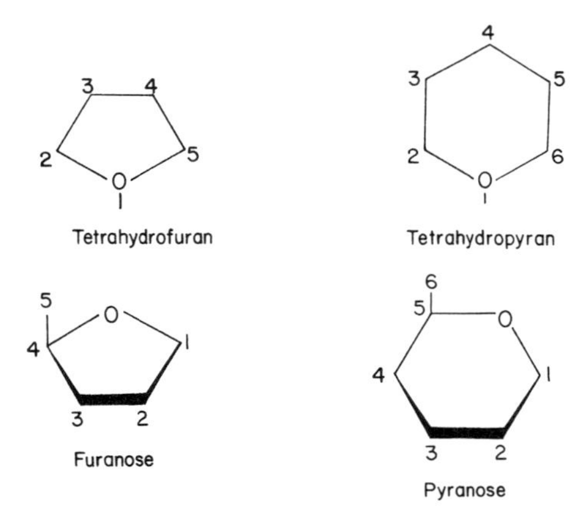

Usually, the two furanose forms are favored kinetically, whereas the two pyranose forms are favored thermodynamically and therefore preponderate at equilibrium. The composition of the equilibrium mixture and the contribution of each of the various forms differ from one sugar to another (see Table I), depending on the instability factors found in each

TABLE I

Composition of Aqueous Solutions of D-Aldoses at Equilibrium

Aldose	% Pyranose			% Furanose[a]		
	α	β	Total	α	β	Total
Allose	16	71	87	(3.5	5	8.5)
Altrose	27	40	67	20	13	33
Glucose	38	62	100	0.1	<0.2	
Mannose	65.5	34.5	100	—	—	—
Gulose	<0.1	78	78	<0.1	22	22
Idose	39	36	75	11	14	25
Galactose	29	64	93	3	4	7
Talose	40	29	69	20	11	31
Ribose	21	59	80	6	14	20
Arabinose	63	34	97	(2.5	2	4.5)
Xylose	36.5	63	99.5			<0.5
Lyxose	70	28	98	1.5	0.5	2

[a] Figures in parentheses were obtained by different authors, and totals therefore do not add up to 100%.

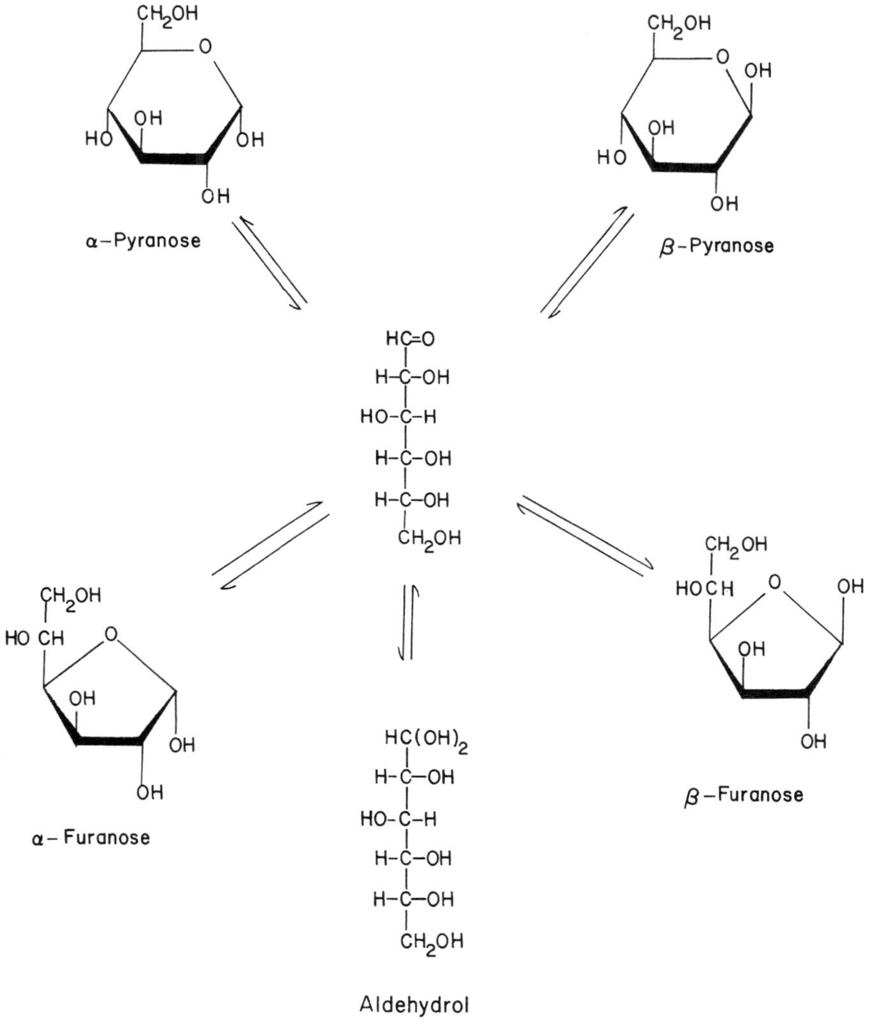

α-Pyranose

β-Pyranose

α-Furanose

β-Furanose

Aldehydrol

tautomer. Thus, for example, because the interaction between *cis* OH groups on adjacent carbon atoms renders this arrangement considerably less desirable in a furanose than a pyranose ring, it was found that the all-*trans* β-D-galactofuranose contributes more to the equilibrium (3%) than β-D-glucofuranose (0.1%).

To study the structure and configuration of one of the aldose forms undergoing rapid equilibration in solution, quick measurements, generally physical ones, are needed. If, on the other hand, a slow chemical method is chosen, the cyclic sugar must first be converted into a derivative, such as a glycoside, that does not undergo equilibration in neutral solution.

A. Anomeric Configuration of Monosaccharides

The anomeric configuration refers to the relative orientation of the substituents attached to C-1 of a cyclic aldose or C-2 of a cyclic 2-ketose. These atoms are achiral in the acyclic forms but become chiral as a result of their conversion into cyclic hemiacetals. Two *anomeric* furanoses and two anomeric pyranoses may be produced from an acyclic monosaccharide, provided the latter possesses the requisite number of atoms (four carbon atoms for a furanose ring and five for a pyranose ring). The configuration of the anomeric center is designated by the Greek letters α and β, which were originally assigned on the basis of optical rotation. Thus, Hudson's rotation rule states that, in the D series, the more dextrorotatory isomer of an anomeric pair will be designated α and the other anomer β. Conversely, in the L series, the more levorotatory member will be designated α and the other anomer β.

Later, as the absolute configuration of a sufficient number of α and β anomers was determined, it became apparent that a relationship exists between the α and β notation as determined by Hudson's rule and the configuration of the first and last chiral centers (the anomeric carbon atom and the center that determines the D and L assignment). This relationship, for free cyclic sugars, may be expressed in one of the following ways.

(a) Using the D and L notation, the first definition states that an anomer, whether in the furanose or pyranose form, is designated α if the anomeric carbon atom has the same (D or L) configurations as the last chiral center (their OH groups point in the same direction in the Fischer projection formula) *prior to cyclization*. The unsuitability of this system is demonstrated by the fact that, in the cyclized product, if the OH group attached to the last chiral center is involved in ring formation, its C–O bond will not be parallel to the anomeric C–OH, but perpendicular to it.

(b) A definition based on the Cahn–Ingold–Prelog system would state that an anomer is designated α if the (R) or (S) signs of the first and last chiral centers are opposite, either (RS) or (SR). If both signs are identical (both R or both S), the anomer is designated β.

(c) Finally, it may be simply stated that, in the D series, if the OH on the anomeric carbon atom is to the right in a Fischer projection or below the plane of a Haworth perspective formula, the isomer is α, and it is β if the OH is to the left or above. Conversely, in the L series, the α anomer has the OH-1 to the left in a Fischer projection or above in a Haworth

α-D-Furanose β-D-Furanose α-L-Furanose β-L-Furanose

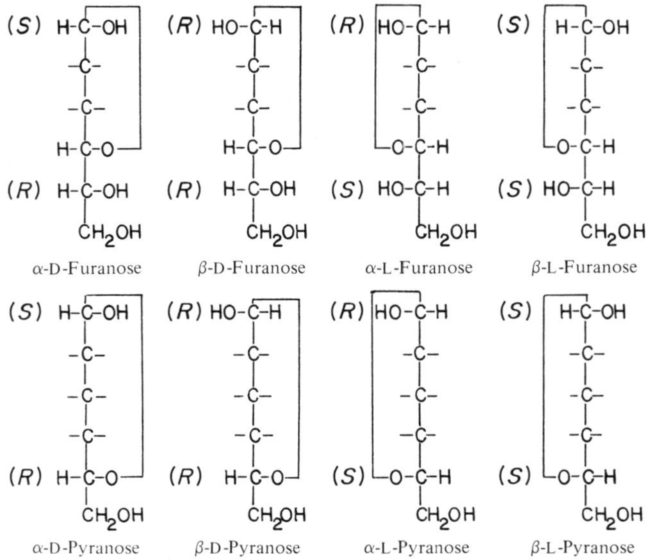

α-D-Pyranose β-D-Pyranose α-L-Pyranose β-L-Pyranose

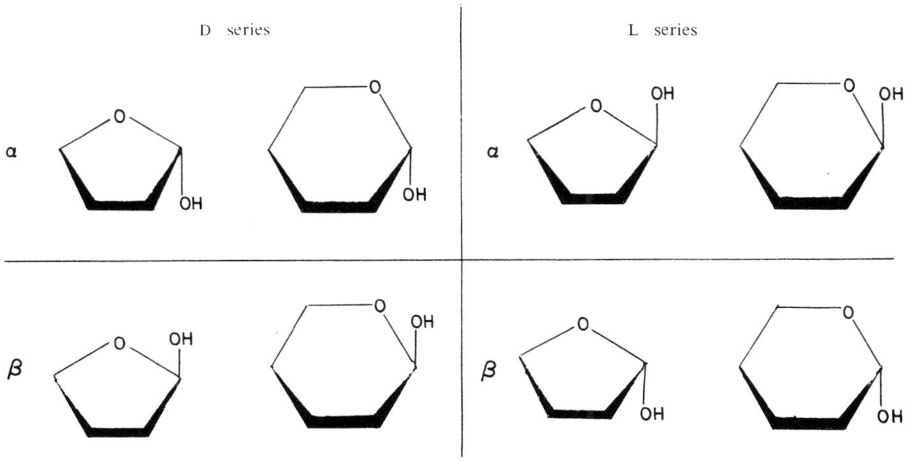

D series L series

formula (see page 41), and the β anomer has this OH to the right or below, respectively.

It should be noted that the foregoing three definitions hold not only for aldopentofuranoses and aldohexopyranoses, where C-1 and the chiral center that determines the D or L notation are both attached to the ring oxygen atom (which, as will be seen later, render their contribution to the observed rotation decisive), but are also valid for higher sugars. This is because a change from the D to the L configuration in any sugar brings about an inversion of all other centers, including the chiral center attached to the ring oxygen atom.

B. Ring Size of Monosaccharides

Although five rings depicted in Fig. 6 are all possible and may be created in some synthetic sugar derivatives, only the furanose and the pyranose rings are found naturally in aldoses.

C. Determination of Ring Size and Anomeric Configuration

In general, physical methods are used to ascertain the ring size and anomeric configuration of monosaccharides, whereas chemical methods are used to determine the ring size of their derivatives.

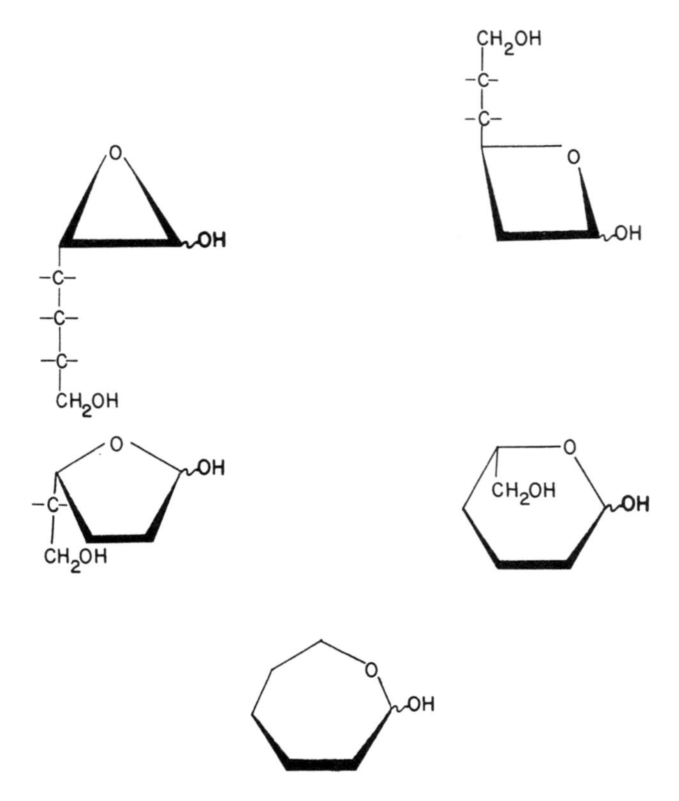

Fig. 6. Possible rings in hexoses.

1. Physical Methods

1. The best and most reliable way to determine, in one set of measurements, the structure, relative configuration, and conformation of an organic compound is by X-ray crystallography or neutron diffraction. Before the advent of computers, these methods were so time-consuming that they were seldom used by organic chemists. The situation has now changed, and X-ray crystallography is quickly becoming a routine analysis. However, it has the drawbacks that (a) it is still a costly analysis (commercial laboratories charge about $1000 per structure determination), (b) it requires that the compound be crystalline (some saccharides are not), and (c) the crystals needed must be of the minimum size that allows the measurements to be carried out on a single crystal. The advantage of this method of analysis is obvious; it not only determines the ring size but affords in one experiment the complete structure, configuration, and conformation of the compound under investigation. Figure 7 shows the ring structure, anomeric configuration, and conformation of methyl α-D-glucopyranoside as deduced by neutron diffraction. It should be emphasized that the favored conformation deduced by X-ray crystallography may differ from that existing in solution as determined by NMR spectroscopy.

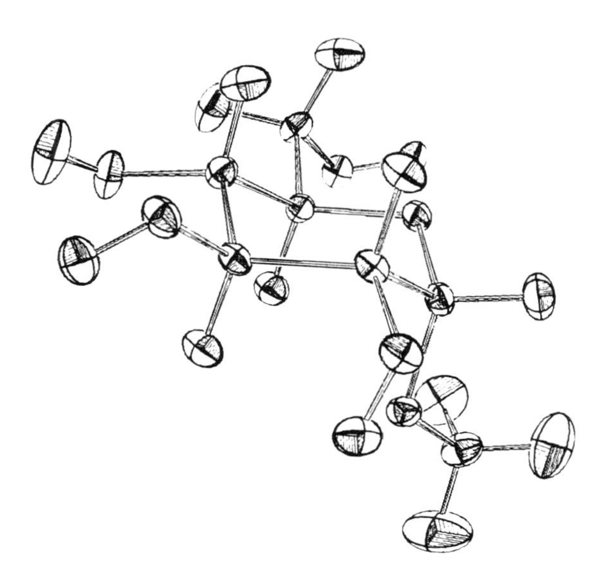

Fig. 7. Neutron diffraction analysis of methyl α-D-glucopyranoside. (Original provided by Professor G. A. Jeffrey.)

2. ^{1}H-Nuclear magnetic resonance spectroscopy constitutes an excellent tool for study of the anomeric configuration of saccharides. The anomeric proton resonance is readily identified at the low-field end of a ^{1}H-NMR spectrum of a monosaccharide, because it is the only proton in the molecule that is attached to a carbon atom bearing two oxygen atoms and hence is significantly less shielded than the other protons. Furthermore, as it is coupled by only one α proton, its splitting is simple, which permits first-order resolution and accurate measurement of the coupling constant. Then, by use of the Karplus curve (see Fig. 8), it is possible to estimate the dihedral angle between the anomeric and adjacent protons (the angle between H-1 and H-2 in an aldose). This will not only establish the anomeric configuration but also give an idea of the ring size and ring

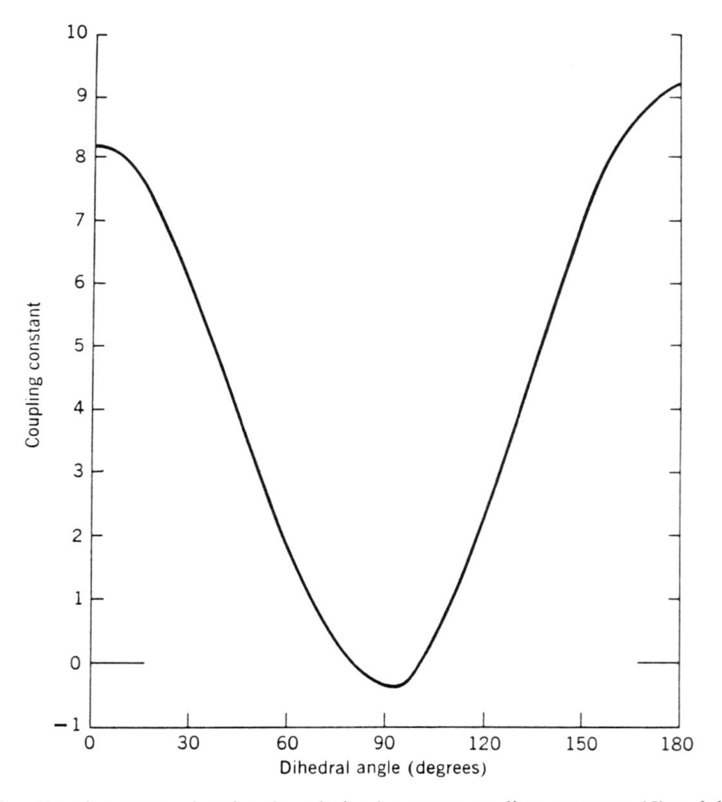

Fig. 8. Karplus curve, showing the relation between coupling constants (J) and dihedral angles ϕ. The J values were determined by using one of the following equations: $J_{H,H'} = 8.5 \cos \phi^2 - 0.28$ (for angles between 0 and 90°), or $J_{H,H'} = 9.5 \cos \phi^2 - 0.28$ (for angles between 90 and 180°).

conformation (see page 45). Then, by using selective decoupling techniques to identify the other protons, it is possible to determine their orientation relative to one another. The ^{1}H-NMR spectrum of a solution of α-D-glucopyranose was shown in Fig. 5.

Because multinuclear magnetic resonance spectrometers having Fourier transform capabilities are now readily available, it is customary to confirm the ^{1}H-NMR results with natural-abundance ^{13}C-nuclear magnetic resonance spectroscopic measurements.

3. In general, optical rotation is the tool of choice in the study of chirality. It is not surprising, therefore, that it has been extensively used in carbohydrate chemistry to ascertain the anomeric configuration of cyclic sugars. Thus, as stated earlier, *Hudson's rotation rule* has been used to determine the anomeric configuration of sugars and their glycosides (see page 32 for a definition of the α and β notation). This rule stipulates that, for an anomeric pair of cyclic sugars or glycosides in the D series, the more dextrorotatory member is the α anomer (and the more levorotatory, the β anomer). Conversely, in the L series, the more levorotatory is the α anomer (and the more dextrorotatory, the β one). It should be noted that in order to use this rule properly, the rotation of *two* enantiomers must be measured, making sure that they really constitute an anomeric pair (they both have the same ring size). Table II shows the rotation of some anomeric aldopyranoses and the (*R*) and (*S*) configurations of their first and last chiral centers (those attached to the ring oxygen atom).

4. The electrical conductivity of solutions provides valuable information on the configuration of carbon atoms bearing hydroxyl groups. Thus, it is possible to determine whether two OH groups attached to adjacent carbon atoms are sufficiently close to one another to form a cyclic borate (as, for example, in a *cis*-1,2-dihydroxy arrangement). Böeseken studied

TABLE II
Optical Rotation of Some Anomeric Aldopyranoses

Aldose	Configuration		Rotation
	C-5	Anomeric C-1	(deg)
α-D-Glucose	(*R*)	(*S*)	+112
β-D-Glucose	(*R*)	(*R*)	+18
α-D-Galactose	(*R*)	(*S*)	+151
β-D-Galactose	(*R*)	(*R*)	+53
α-D-Mannose	(*R*)	(*S*)	+29
β-D-Mannose	(*R*)	(*R*)	+17
α-L-Arabinose	(*S*)	(*R*)	+40
β-L-Arabinose	(*S*)	(*S*)	+191

the conductivity of a solution of α-D-glucopyranose in a boric acid buffer and found that it decreased with time, whereas the conductivity of a similar solution of β-D-glucose increased with time, until both reached the same specific conductivity. He concluded that, in the α form, the 1- and 2-hydroxyl groups must be cis to one another to enable them to come close enough to form a cyclic borate (which conducts electricity). Although Böeseken was right in assigning the 1,2-cis configuration to α-D-glucose, careful examination of models reveals that neither α-D-glucopyranose nor its β-D form in a chair conformation has its OH groups as close as in α-D-glucofuranose. The anomeric configuration he deduced was that of the latter compound.

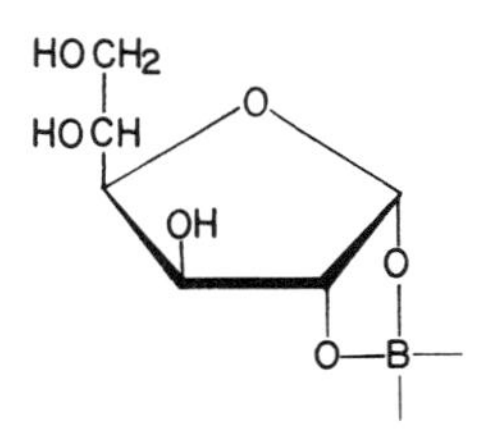

Borate complex

2. Chemical Methods

In order to use a chemical reaction to ascertain the structure of a compound undergoing equilibration, either a reaction that proceeds faster than the equilibration must be used, or the compound must be converted into a form that does not undergo equilibration before its structure can be studied. The following three methods have been used to determine the ring size of sugars and their glycosides.

1. Rapid oxidation of the furanose and the pyranose forms of D-glucose with bromine–water would be expected to yield, in the first case, a 1,4-lactone and, in the second case, a 1,5-lactone. Actually, it was found by Isbell that oxidation of α-D-glucose yields D-glucono-1,5-lactone, indicating that the ring is six-membered.

The two remaining methods in this section involve converting the cyclic form of a monosaccharide into an acetal (to prevent the ring from reopening) and then subjecting the acetal to a chemical test. The acetals obtained by treating a monosaccharide with alcohols under catalysis are called *glycosides,* and the specific ones are named after the parent sugar—thus, glucosides, mannosides, etc. A large amount of the structural work on the cyclic forms of monosaccharides was conducted on methyl glycosides because, unlike the free hemiacetal forms, these compounds do not equilibrate in neutral solutions. To prepare them, D-glu-

α-D-Glucofuranose → D-Glucono-1,4-lactone

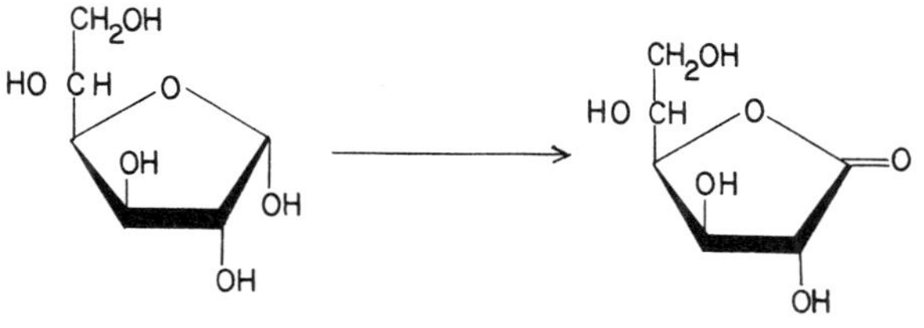

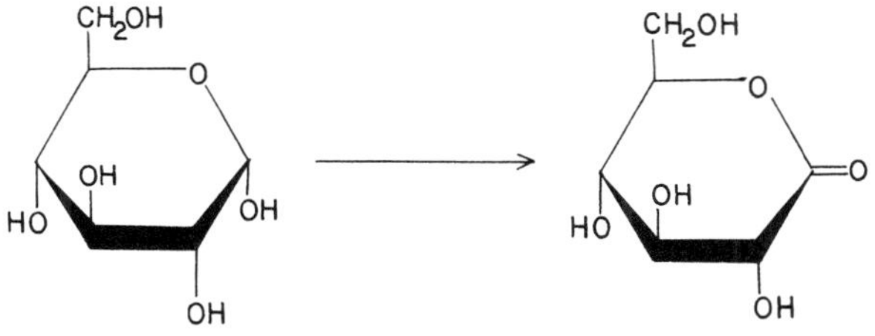

α-D-Glucopyranose → D-Glucono-1,5-lactone

cose, for example, is refluxed with methanol in the presence of an acid catalyst, whereby the kinetically favored methyl α- and methyl β-D-glucofuranosides are formed first, but as the reaction proceeds the thermodynamically favored methyl α- and methyl β-D-glucopyranosides become the preponderant products. Accordingly, if there is interest in obtaining the furanosides, a short reaction period is prescribed, and if the pyranosides are desired, the reaction time is prolonged to permit the equilibrium to be reached.

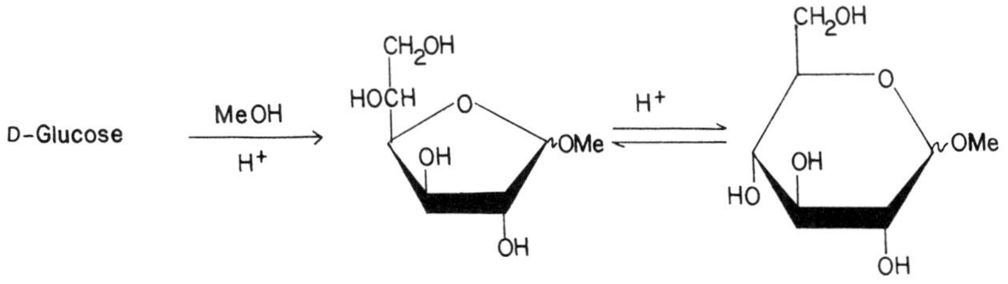

D-Glucose $\xrightarrow[H^+]{MeOH}$ Methyl D-glucofuranoside (kinetically favored) $\overset{H^+}{\rightleftharpoons}$ Methyl D-glucopyranoside (thermodynamically favored)

2. The first use of chemical labeling in elucidating the structure of an organic compound was made by Haworth in his study of the ring size of methyl α- and β-D-glucopyranoside. Recognizing the difference between an ether methoxyl group, which is relatively stable toward acid- and base-

catalyzed hydrolysis, and a glycosidic methoxyl group, which is an acetal and is susceptible to acid hydrolysis, he designed a brilliant experiment to determine the size of the ring in a glycoside. He labeled all of the free hydroxyl groups in methyl α- and β-D-glucopyranoside by etherification with dimethyl sulfate in alkali and then removed the glycosidic methoxyl group by acid hydrolysis, which left carbon atom 1 unlabeled. Because, at that time, no authentic methylated sugars were available for comparison with the resulting product, he broke the saccharide chain at the site of the unprotected hydroxyl group by vigorous oxidation and obtained a methylated dicarboxylic acid (trimethoxyglutaric acid) having the same number of carbon atoms in its chain as the sugar ring. Currently there exist, in various research centers, complete libraries of authentic sugars methylated in various positions that may be used for comparison with a reaction product in conjunction with such techniques as thin-layer chromatography, gas chromatography, or liquid chromatography. These reference compounds are particularly useful in determining the ring size and position of linkage in oligo- and polysaccharides.

Methyl α-D-glucopyranoside $\xrightarrow[OH^-]{Me_2SO_4}$ Methyl 2,3,4,6-tetra-*O*-methyl-α-D-glucopyranoside $\xrightarrow{H^+}$ 2,3,4,6-Tetra-*O*-methyl-D-glucopyranose → Trimethoxyglutaric acid

3. Another method for determining the ring size of glycosides is periodate oxidation, which proceeds stoichiometrically and is a measure of the number of adjacent free hydroxyl groups. In this procedure, developed by Malaprade, the moles of periodic acid (or sodium metaperiodate) consumed by, and the moles of formaldehyde and formic acid produced during the oxidation of, a known weight of the saccharide under investigation are determined. It is known that 1 mole of this oxidant is reduced (consumed) when two adjacent OH groups are oxidized, with cleavage of the C–C bond joining them, to yield two aldehyde groups. Specifically, a terminal (primary) hydroxyl group yields formaldehyde on oxidation; a secondary hydroxyl group yields another aldehyde (which is not usually estimated) or formic acid if the secondary hydroxyl group is flanked on both sides by hydroxyl groups (i.e., is oxidized twice). By ascertaining the outcome of the oxidation, it is possible to determine the number and

Periodate oxidation of the possible methyl α-D-glucosides

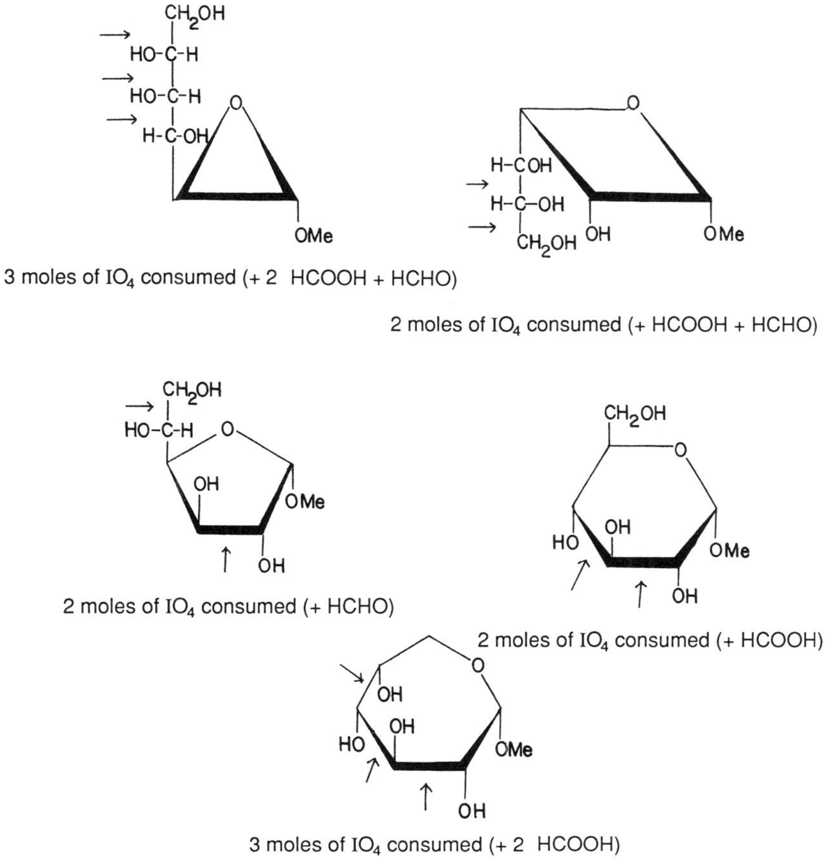

3 moles of IO_4 consumed (+ 2 HCOOH + HCHO)

2 moles of IO_4 consumed (+ HCOOH + HCHO)

2 moles of IO_4 consumed (+ HCHO)

2 moles of IO_4 consumed (+ HCOOH)

3 moles of IO_4 consumed (+ 2 HCOOH)

type (primary or secondary) of adjacent hydroxyl groups present in a molecule.

To illustrate this, we will now calculate the moles of periodate consumed and the moles of formaldehyde and formic acid produced in the oxidation of the theoretically possible ring types of a methyl glycoside, i.e., the various methyl glycosides having three-, four-, five-, six-, and seven-membered rings. The calculated figures will then be compared with the experimental results to determine the ring size with which they agree. Periodate oxidation of both methyl α- and β-D-glucopyranoside was found to result in the consumption of 2 moles of oxidant and the liberation of 1 mole of formic acid (no formaldehyde was produced), which is in agreement with the results calculated for a six-membered ring. On the other hand, oxidation of the methyl D-glucofuranosides resulted in the consumption of 2 moles of periodate and the liberation of 1 mole of formaldehyde (no formic acid was produced), which corresponds to the calculated values expected from a five-membered ring. It should be noted that periodate oxidation is not useful in the study of the free (reducing) sugars, because the hemiacetal ring in these compounds can be opened, thus giving the results expected from the acyclic form.

D. The Haworth Perspective Formula

The common method of depicting monosaccharides in their cyclic forms, without assigning a specific conformation to the ring, is by using the Haworth perspective formula. The ring, a pentagon having 108° angles or a hexagon with 120° angles, was originally represented as viewed by an observer situated above its plane. Today, it is represented as seen by an observer situated at an angle of about 60° above the plane of the ring and, as a precaution against optical illusions that could shed doubt on the side closer to the viewer, the ring bonds nearest to the observer are thickened to give a sense of perspective.

Although a ring may be turned around its center, by the rules of carbohydrate nomenclature it is oriented in such a manner that C-1 is to the right and the ring oxygen atom is farthest from the viewer. The only chemical symbol used in the ring is that of oxygen; carbon atoms are represented by angles and have vertical bonds connecting them to H (optional), OH, C, or some other atom (e.g., N or S).

The conversion of a Fischer formula into a Haworth formula is best done by first drawing the Fischer formula (for example, that of D-glucose, shown next). Then, for an α-D-pyranoside, a nucleophilic attack is initiated by O-5 on the 1-carbonyl in such a manner that the chiral center created acquires the (*S*) configuration. To do this, its OH must point in the same direction (to the right) as the OH on C-5. Then C-5 is turned on its axis in order to bring O-5 behind the ring, and by so doing, H-5 is moved from left to right and C-6 from behind to the left. The final step is to lay the chain on its side by moving the top of the molecule to the right. Now the groups or atoms that were to the right in the Fischer projection

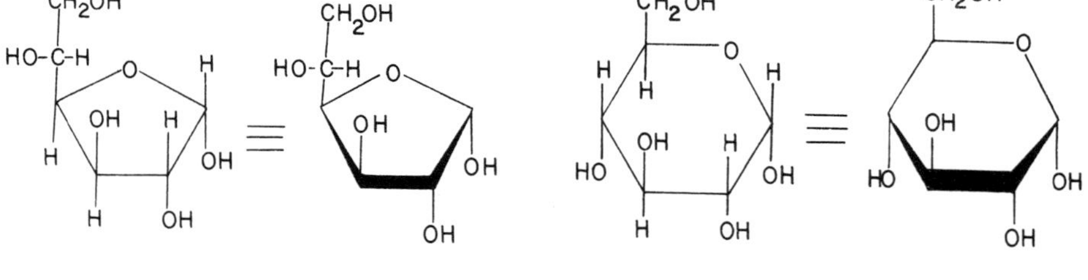

are below the ring, and those to the left are above the ring. The bond distances and angles are equalized and the molecule is drawn as viewed.

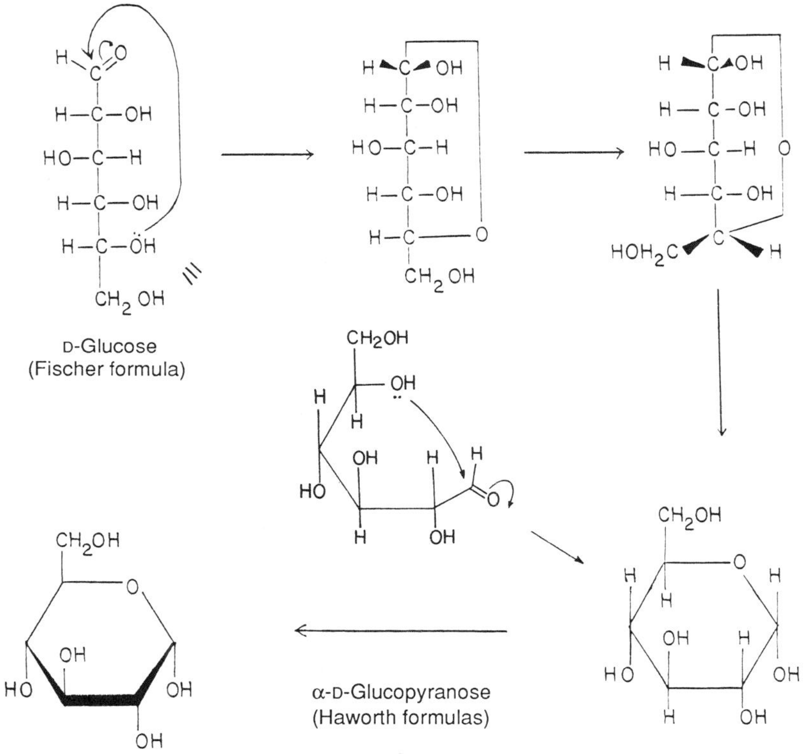

D-Glucose
(Fischer formula)

α-D-Glucopyranose
(Haworth formulas)

The following are some useful suggestions regarding conversions from Fischer to Haworth structures for an aldose.

(a) The side chain (contains C-5 in a furanose or C-6 in a pyranose) is up when the configuration of the ring carbon atom to which it is attached is (*R*) (originally had an OH to the right) and down for the (*S*) configuration.

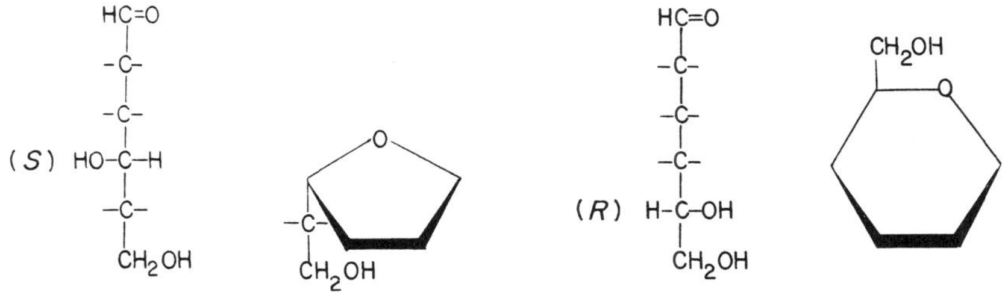

(b) If the side chain is *down*, the position of the substituents around any chiral center in the chain is the same as it is in the Fischer formula. On the

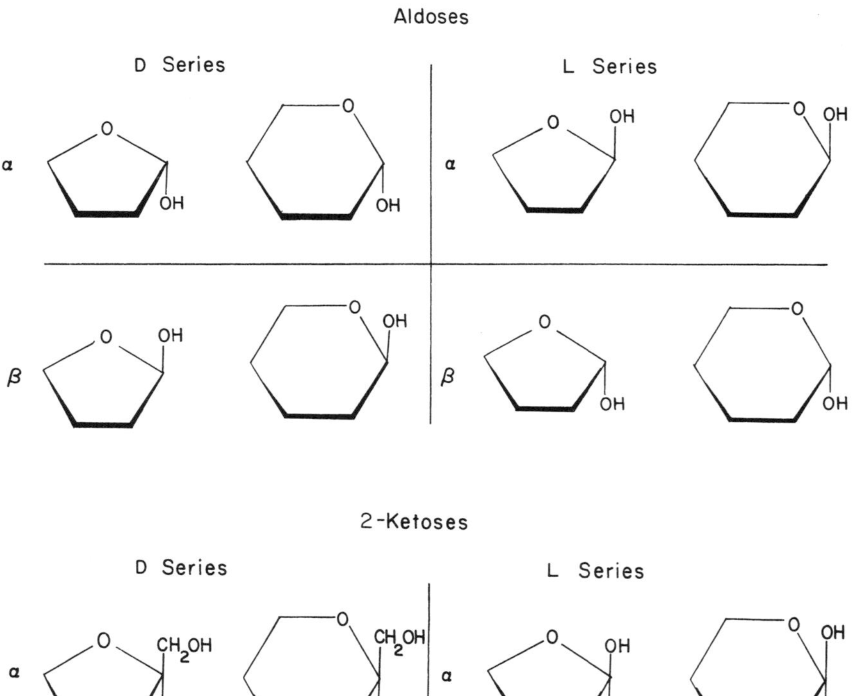

other hand, if the chain is *up*, the position of a substituent attached to a chiral center in the chain will seem to be the reverse of what it was in the Fischer formula. This is because in the latter the last carbon atom is down, whereas now this carbon is up.

(c) For D-aldoses, the α anomer has C-1 in the (S) configuration and the OH is down, and the β anomer has C-1 in the (R) configuration and the OH up. For L-aldoses, the α anomer has C-1, (R), and the OH up, and the β anomer has C-1, (S), and the OH down.

(d) Assigning an α or β configuration to a 2-ketose, or representing a ketose anomer using a Haworth formula, may seem confusing. The easiest way is simply to draw the anomeric OH group in the same direction (up or down) as in the corresponding aldose anomer and then replace the anomeric hydrogen atom of the latter with CH_2OH. When this is done, the β anomer will have acquired the same (R) or (S) configuration as the last chiral center (C-5 of an aldohexose), and the α anomer will have acquired the opposite configuration.

IV. CONFORMATION OF MONOSACCHARIDES

The next stage in the structural investigation of a monosaccharide is the determination of the conformation of the ring or, if the monosaccharide is acyclic or has a long side chain, the conformation of the chain.

Furanose and pyranose rings can exist in a number of interconvertible conformers that differ in stability from one form to another. It will be useful at this stage to enumerate the various furanose and pyranose conformers and to discuss in some detail the methods for determining the conformation of pyranoses.

A. Conformation of the Furanose Ring

The principal conformers of the furanose ring are the *envelope* (E) and the *twist* (T) forms. A particular conformation is designated by the same method for furanoses and pyranoses. Here, the letter used to designate the form (for example, E or T) is preceded by the number of the ring atom situated above the plane of the ring and is followed by the number of the ring atom below the plane of the ring; a ring oxygen atom is designated O. Again, to emphasize the locant position in relation to the ring, the former numbers are always superscripted and the latter subscripted, e.g., ${}^{2}T_{3}$. It may be noted that, as the envelope forms have only one atom above or below the ring, the letter E is accompanied by one number only. The two envelope forms ${}^{O}E$ and E_{O} and the twist form ${}^{2}T_{3}$ are depicted.

Envelope forms

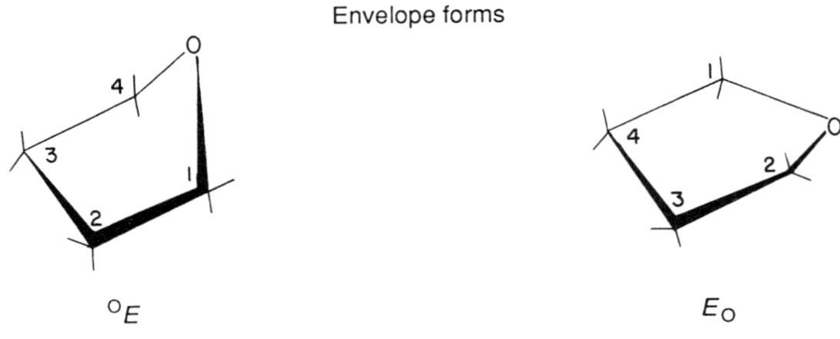

^{O}E E_{O}

Possible envelope forms (10): ^{1}E; E_{1}; ^{2}E; E_{2}; ^{3}E; E_{3}; ^{4}E; E_{4}; ^{O}E; E_{O}

The most stable conformers of the furanose ring are the envelope and twist forms; these can exist in 10 arrangements each. Because of the low energy barriers between the E and T conformers, a sugar in the furanose form is thought to undergo rapid interconversions between them. The slightly more favored twist conformer would rapidly pass through an

Twist forms

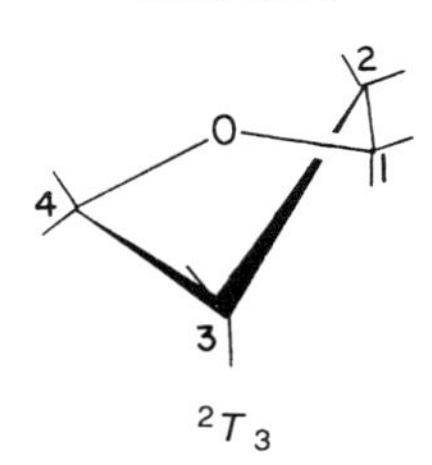

$^{2}T_{3}$

Possible twist forms (10): $^{O}T_{1}$; $^{1}T_{O}$; $^{1}T_{2}$; $^{2}T_{1}$; $^{2}T_{3}$; $^{3}T_{2}$; $^{3}T_{4}$; $^{4}T_{3}$; $^{O}T_{4}$; $^{4}T_{O}$

envelope conformation (which is less favored because it possesses two eclipsed carbon atoms) to go to the next twist form. Because interaction between two eclipsed carbon atoms is greater than that between a carbon atom and an oxygen atom, the ring oxygen atom tends to occupy a position along the plane of the ring and to leave the puckering to carbon

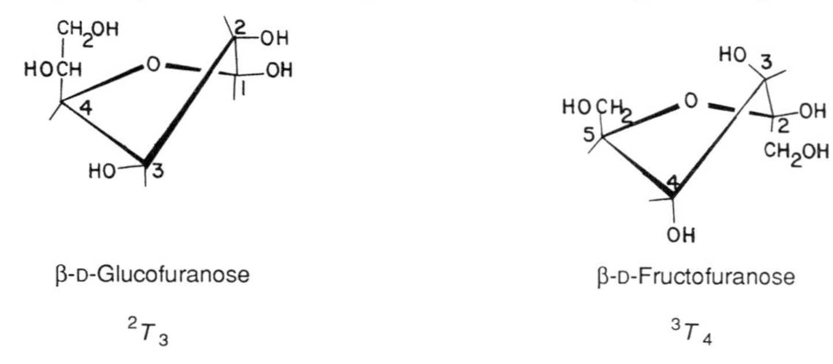

β-D-Glucofuranose β-D-Fructofuranose

$^{2}T_{3}$ $^{3}T_{4}$

atoms. β-D-Glucofuranose and β-D-fructofuranose are depicted in the $^{2}T_{3}$ and $^{3}T_{4}$ conformations.

B. Conformation of the Pyranose Ring

Recent measurements have revealed that the bond angles in the pyranose ring are 111° for carbon, only slightly greater than the expected value of 109.5°, and 113° for oxygen, considerably greater than originally presumed. Accordingly, the shape of the pyranose ring, although somewhat flatter, resembles to a great extent that of the cyclohexane ring. The numbers of recognized forms of the pyranose ring are two *chair* (C), six *boat* (B), four *half-chair* (H), six *skew* (S), and six *sofa* forms. To designate each of these forms, the number(s) of the ring atom(s) lying above the plane of the pyranose ring is (are) put before the letter designating the form (C, B, H, S, etc.), and the number(s) of the ring atom(s) lying below the plane is (are) put after the letter; a ring oxygen atom is designated O. To emphasize the location of the atoms vis-à-vis the plane of the ring, the former numbers are superscripts and the latter are subscripts. Thus, $^{4}C_{1}$ and $^{1}C_{4}$ designate the two chair forms, which are depicted. Also depicted are two boat forms, a half-chair, a skew, and a sofa form. It may be noted that the sofa forms, which have only one atom above or below the ring, are designated with only one numeral or symbol (O). The energy barriers between the conformers of the pyranose ring compared to those in cyclohexane were measured by Horton, using NMR spectroscopy.

Chair forms

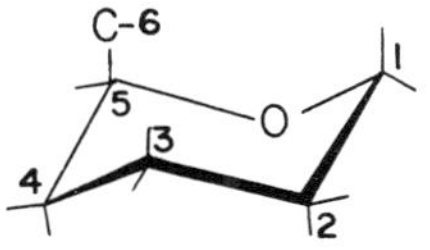

$^{1}C_{4}$ 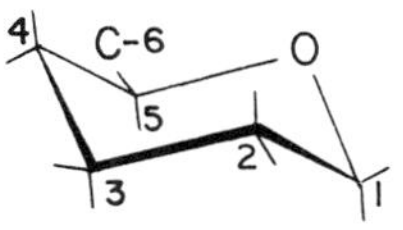 $^{4}C_{1}$

Possible chair forms (2): $^{1}C_{4}$ and $^{4}C_{1}$

Boat forms

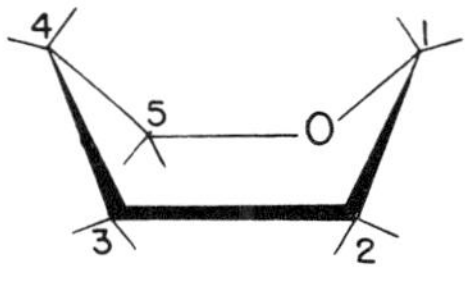

$^{1,4}B$ 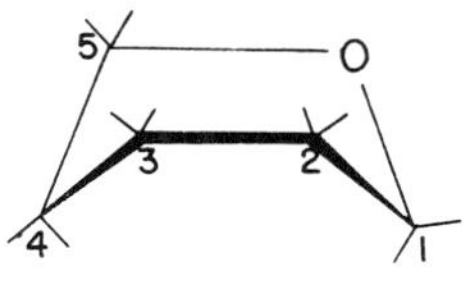 $B_{1,4}$

Possible boat forms (6): $^{1,4}B$; $B_{1,4}$; $^{2,5}B$; $B_{2,5}$; $^{O,3}B$; $B_{O,3}$

Skew forms

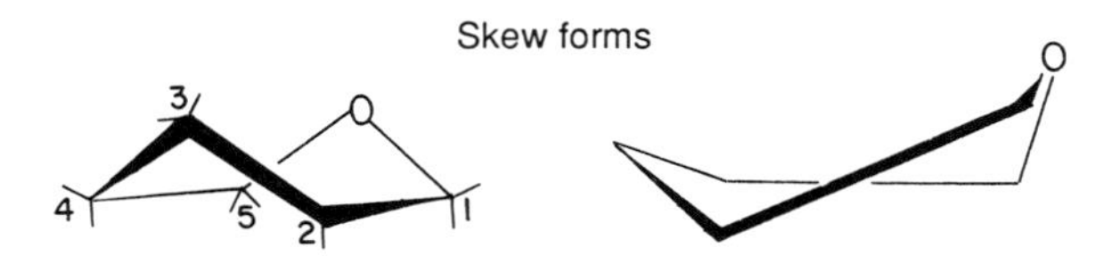

Possible skew forms for aldopyranoses (6): ${}^{1}S_{5}$; ${}^{5}S_{1}$; ${}^{2}S_{O}$; ${}^{O}S_{2}$; ${}^{1}S_{3}$; ${}^{3}S_{1}$

Half-chair forms

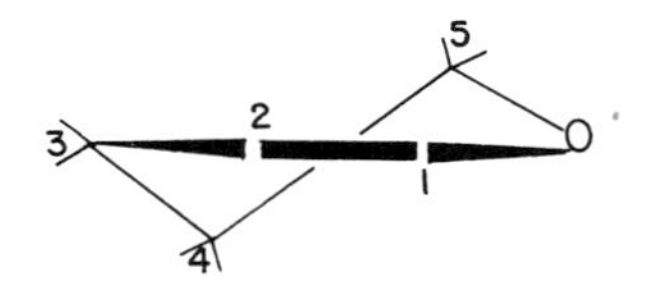

${}^{5}H_{4}$

Possible half-chair forms for aldopyranoses (12): ${}^{O}H_{1}$; ${}^{1}H_{O}$; ${}^{1}H_{2}$; ${}^{2}H_{1}$; ${}^{2}H_{3}$; ${}^{3}H_{2}$; ${}^{3}H_{4}$; ${}^{4}H_{3}$; ${}^{4}H_{5}$; ${}^{5}H_{4}$; ${}^{5}H_{O}$; ${}^{O}H_{5}$

Sofa forms

Possible sofa forms (6): 1; 2; 3; 4; 5; O

C. Determination of Ring Conformation

1. X-Ray Crystallography and NMR Spectroscopy

Although it is possible to predict which conformer of a given ring (furanose or pyranose) will be more stable, it is often desirable to verify this experimentally, by either X-ray crystallography or NMR spectroscopy. Both these methods of analysis have already been discussed and will not be examined again. Suffice it to say that single-crystal X-ray crystallography provides the exact location of all nonhydrogen atoms in a crystal lattice and shows the exact conformation of the ring. The conformation of a saccharide in a particular solvent can also be determined from the ^{1}H-nuclear magnetic resonance spectrum. The relative orientation of hydrogen atoms attached to successive pairs of carbon atoms, for example, the angle between H-1 and H-2, H-2 and H-3, H-3 and H-4, and H-4 and H-5, may be determined from their respective coupling constants using the Karplus equation. Then, by summing up the data, the conformation of the whole molecule may be determined. From measurements made at different temperatures, it is also possible to determine the energy barriers between conformers. It must be emphasized that the conformation deduced by X-ray crystallography reflects the situation in the solid state; and that deduced from NMR measurements is the conformation in the solvent in which the measurement was taken.

2. Optical Properties of Heavy-Metal Complexes

Because saccharides possess adjacent hydroxyl groups which can complex with heavy metals, their ring conformation can be determined by ascertaining which pair of hydroxyl groups is involved in chelate formation. This approach was used by Reeves, who found that chelation with copperammonium salts resulted in an enhancement of optical rotation by several orders of magnitude. To determine which hydroxyl groups were involved in complexation, methoxyl groups were introduced on C-2, C-4, or C-3 and used to block the formation of 1,2, or 3,4, or both complexes, respectively. It was possible then to determine which hydroxyl groups were sufficiently close to form the metal complexes. This method of analysis is now of historical interest only, because it can give misleading results, particularly when the heat of formation of the complex exceeds the energy barrier that separates one conformer from another.

3. Theoretical Calculations for Pyranose Rings

Of the various conformations possible for the pyranose ring, the two chair conformations are so much more stable than the others that, unless overwhelming reasons dictate otherwise, the analysis should be restricted to the choice between the $^4C_1$ and $^1C_4$ conformers. Four methods have been used to predict which of these is the stable conformer.

a. *Instability Factors.* Of historical interest is a method developed by Reeves, who assigned numerical values to the following instability factors and regarded the form having the smaller numerical total as the more stable conformer.

(1) The most important instability factor in the pyranose ring (with a value of 2.5) is a situation called delta 2. This arises when the OH on C-2 bisects the angle between the ring oxygen atom and the oxygen atom of the anomeric hydroxyl group, forming an isosceles triangle having one oxygen atom at each corner.

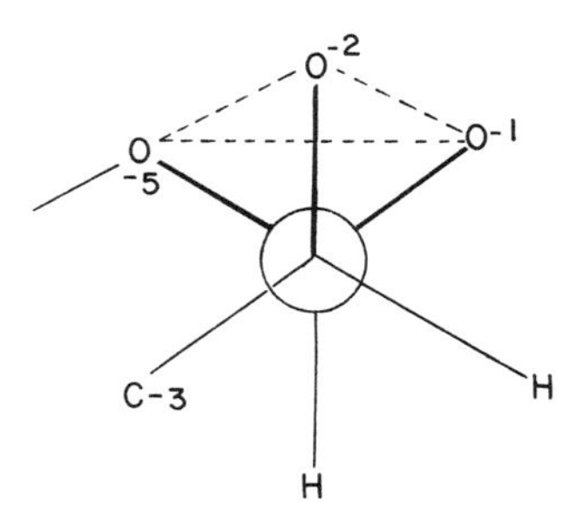

delta 2 situation

(2) An instability factor of 2.0 is given to an axial CH_2OH.

(3) An axial hydroxyl group on the ring has an instability factor of 1.0 unit.

(4) The 1,3 interaction between an axial CH_2OH and an axial OH group on the same side of the ring produces an unfavorable situation with an additional value of 2.5.

1,3 Interaction

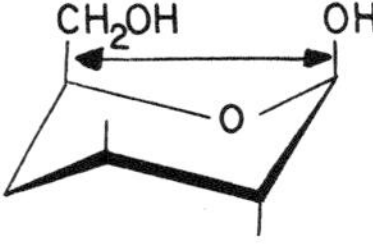

b. *Nonbonded Interaction Energies.* This method was developed by Angyal, who based his calculations on two assumptions: (a) that the geometry of the pyranose ring is the same as that of the cyclohexane ring and (b) that the free energies of the conformers are additive functions, which are independent of one another. The following interaction energies have been used to ascertain the stability of the conformers.

Axial interactions (kJ/mole)	Gauche interactions (kJ/mole)
H and O = 1.7	O and O = 2.1
H and CH_2OH = 3.8	O and CH_2OH = 2.1
O and O = 6.3	
O and CH_2OH = 10.5	

c. *Thermodynamic Calculations of the Free Energy.* Empirical calculations of the conformational free energy by summation of the unfavorable interactions have been used successfully in conformational analysis to determine the more stable conformer of a pyranose in a chair form. Two types of interactions are differentiated, those between two gauche 1,2 substituents [either equatorial–equatorial (*ee*) or axial–equatorial (*ae*)] and those between two syn-diaxial 1,3 groups.

The following are the free energies assigned to some of these groups.

	Free energy (kJ)
Gauche 1,2 interactions	
O-1-*e*–O-*e*	2.3
O-1-*e*–O-*a*	4.2
O-*e*–O-*e* or *a*	1.5
C-*e*–O-*e* or *a*	1.9
Axial–axial 1,3 interactions	
O–O	6.3
O–C	10.5
O–H	1.9
C–H	3.8
Anomeric effect (depends on nature of OR-1 and OH-2)	1.3

By using these values, it was calculated (see below) that the $^4C_1$ form of β-D-glucopyranose has a conformational energy of 8.5, considerably lower than that of the $^1C_4$ form, which has a value of 33.6, thus leaving no doubt that the first is the more stable conformer.

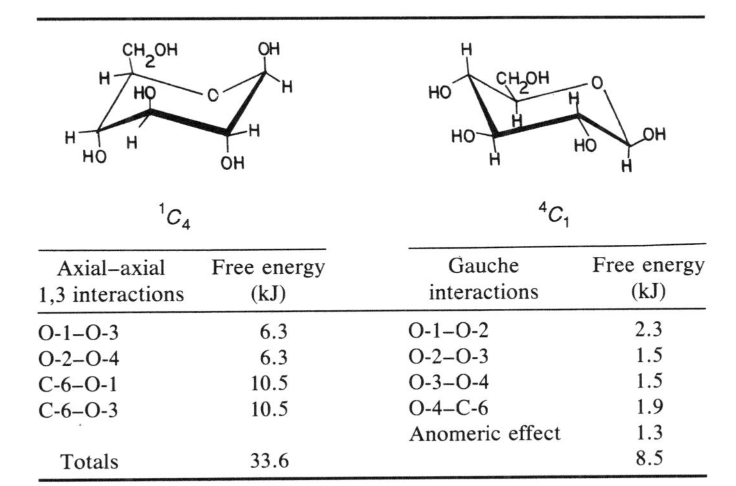

$^1C_4$ $^4C_1$

Axial–axial 1,3 interactions	Free energy (kJ)
O-1–O-3	6.3
O-2–O-4	6.3
C-6–O-1	10.5
C-6–O-3	10.5
Totals	33.6

Gauche interactions	Free energy (kJ)
O-1–O-2	2.3
O-2–O-3	1.5
O-3–O-4	1.5
O-4–C-6	1.9
Anomeric effect	1.3
	8.5

Table III shows the conformation of aldopyranoses in aqueous solutions as obtained from thermodynamic calculations and from interaction energies.

4. Anomeric Effect

Lemieux coined the term anomeric effect to describe the singular behavior of pyranose rings, which differs from what is observed with cyclohexane derivatives. Thus, whereas the equatorial orientation is always preferred over the axial one in substituted cyclohexanes, axial electronegative substituents at the anomeric position are preferred over equatorial ones in pyranose rings. The anomeric effect has been used to explain a

TABLE III

Conformation of D-Aldopyranoses in Aqueous Solutions

	Found by NMR	Found by thermodynamic calculation	Interaction energies (kJ/mole)	
			$^4C_1$	$^1C_4$
Aldohexoses				
α-D-Allose	$^4C_1$	$^4C_1$	16.33	22.40
β-D-Allose	$^4C_1$	$^4C_1$	12.35	25.33
α-D-Altrose	$^4C_1$, $^1C_4$	$^4C_1$, $^1C_4$	15.28	16.12
β-D-Altrose	$^4C_1$	$^4C_1$	14.03	22.40
α-D-Galactose	$^4C_1$	$^4C_1$	11.93	26.38
β-D-Galactose	$^4C_1$	$^4C_1$	10.47	32.45
α-D-Glucose	$^4C_1$	$^4C_1$	10.05	27.84
β-D-Glucose	$^4C_1$	$^4C_1$	8.58	33.49
α-D-Gulose	$^4C_1$	$^4C_1$	16.75	19.89
β-D-Gulose	$^4C_1$	$^4C_1$	12.77	22.82
α-D-Idose	$^4C_1$, $^1C_4$	$^4C_1$, $^1C_4$	18.21	16.12
β-D-Idose	$^4C_1$	$^4C_1$	16.96	22.40
α-D-Mannose	$^4C_1$	$^4C_1$	10.47	23.24
β-D-Mannose	$^4C_1$	$^4C_1$	12.35	32.03
α-D-Talose	$^4C_1$	$^4C_1$	14.86	24.70
β-D-Talose	$^4C_1$	$^4C_1$	16.75	33.49
Aldopentoses				
α-D-Arabinose	$^1C_4$	$^1C_4$	13.40	8.58
β-D-Arabinose	$^4C_1$, $^1C_4$	$^4C_1$, $^1C_4$	12.14	10.05
α-D-Lyxose	$^4C_1$, $^1C_4$	$^4C_1$, $^1C_4$	8.58	10.89
β-D-Lyxose	$^4C_1$	$^4C_1$	10.47	14.86
α-D-Ribose	$^4C_1$, $^1C_4$	$^4C_1$, $^1C_4$	14.44	14.86
β-D-Ribose	$^4C_1$, $^1C_4$	$^4C_1$, $^1C_4$	10.47	12.98
α-D-Xylose	$^4C_1$	$^4C_1$	8.16	15.07
β-D-Xylose	$^4C_1$	$^4C_1$	6.70	16.33

multitude of phenomena that were noted by previous observers but could not be adequately explained. For example, it was known that all the D-glycopyranosyl halides had α configurations (their halogen was axial), irrespective of whether they were obtained from peracetylated α- or β-D-glycopyranoses. Similarly, Pacsu had shown as early as 1928 that methyl tetra-*O*-acetyl-β-D-glucopyranoside could be converted in more than 90% yield into the α anomer with $TiCl_4$ (and not the reverse). Note that this is a much more selective reaction than anomerization.

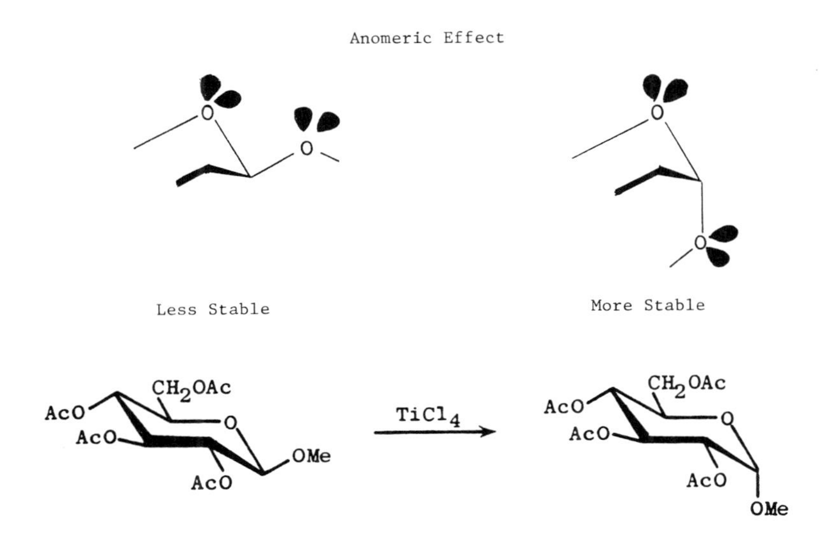

The anomeric effect has been interpreted in terms of polar interactions between the negative group attached to the anomeric position (for example, the halogen atom of a glycosyl halide) and the nonbonded electron pair of the ring oxygen. Newman projections will reveal that an equatorially oriented halogen will be gauche to both electron lobes of the ring oxygen, whereas an axially oriented halogen will be gauche to only one lobe and antiperiplanar to the other.

Another interpretation of the anomeric effect is based on highest occupied molecular orbital–lowest unoccupied molecular orbital (HOMO–LUMO) interactions. Thus, an overlap between the highest occupied molecular orbital of the ring oxygen (the *p* orbital of one of its nonbonding electrons) and a suitably located lowest unoccupied molecular orbital (an antibonding σ^* orbital) of the anomeric carbon will increase the electron density of the group located in an antiperiplanar position relative to the oxygen HOMO. If this group is electronegative it will be stabilized, and if it is a hydrogen it will be destabilized. Accordingly, in the case of α-D-glucopyranosyl chloride, the overlap will be with the σ^* of the C–Cl bond, which will increase the electron density of the electronegative chlorine and stabilize it. In the case of the β-D anomer, the overlap will be with the antibonding orbital of the C–H bond, which will increase the electron density on the hydrogen and produce a destabilizing effect. This destabilizing effect may explain an observation made by H. Isbell in the 1920s, that β-D-glucopyranose (with an axial anomeric hydrogen) reacts much faster with bromine to give the δ lactone than the more stable α anomer.

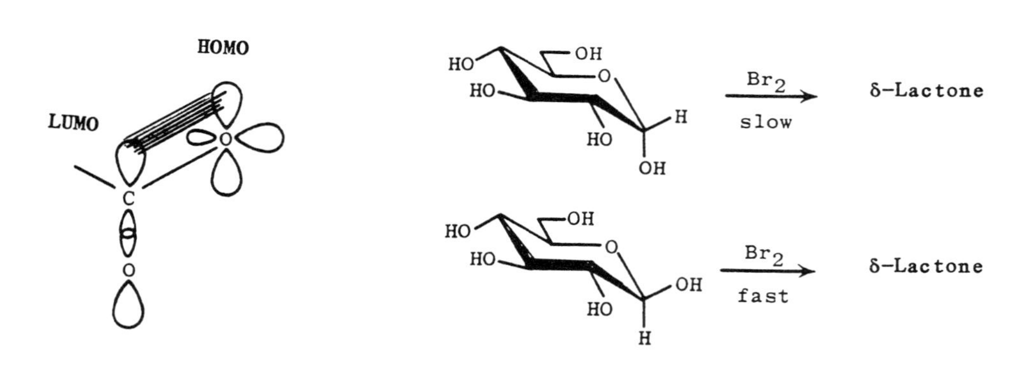

It should be noted that although manifestations of the anomeric effect are stronger in nonpolar solvents than in polar solvents such as water, they are still felt in aqueous solutions. Thus, from the nonbonded interaction energy discussed in the previous section, the free energy of α-D-glucopyranose should be 3.8 kJ/mole higher than that of the β-D anomer (because of the axial H-1 and axial O-1 interaction). In fact, the equilibrium solution contains 36% α and 64% β, which corresponds to a free-energy difference of 1.5 kJ/mole. The 2.3 kJ/mole difference is attributed to the anomeric effect in water.

It is evident that if the glycosidic bond is attached to a group less electronegative than hydrogen, the equatorial anomer will be favored. This reverse anomeric effect was observed by R. U. Lemieux [*Can. J. Chem.* **43,** 2205 (1965); **46,** 1453 (1968)] and by C. Schuerch [*J. Am. Chem. Soc.* **95,** 1333 (1973)] and was explained by S. David [*J. Am. Chem. Soc.* **95,** 3806 (1975)].

Along similar lines of reasoning, one would expect that an exoanomeric effect would favor the all-gauche conformation for methyl α-D-glycopyranosides and not the zigzag all-trans conformation that is prevalent in polymethylenes. This was confirmed by X-ray and neutron diffraction analysis of single crystals of different glucosides and by calculations using dimethoxymethane as a model [see G. A. Jeffey, J. A. Pople, J. S. Binkley, and S. Vishveshvara, *J. Am. Chem. Soc.* **100,** 373 (1978)].

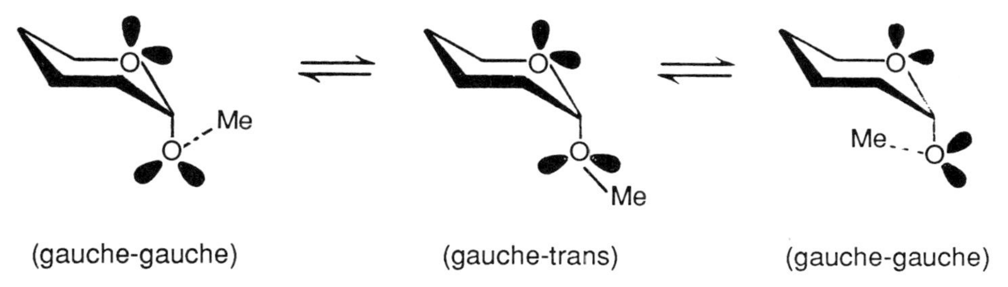

(gauche-gauche) (gauche-trans) (gauche-gauche)

D. Side-Chain Conformation

Although free sugars having four or more carbon atoms exist mainly in cyclic forms, some of their derivatives, for example, dithioacetals, are exclusively acyclic. Furthermore, higher sugars that exist in cyclic

Galactitol (zigzag conformation)

Xylitol (sickle conformation)

forms possess an exocyclic side chain whose conformation may influence the physical and chemical properties of the whole molecule. For these reasons, and to complete the structure elucidation for monosaccharides and their derivatives, it is important to determine the conformation of a carbon chain if a sufficiently long one exists in the molecule.

Unlike hydrocarbon chains, which exist preponderantly in a planar zigzag conformation with the bulkier carbon atoms assuming antiperiplanar positions, the conformation of a hydroxyalkyl chain is complicated by the strong repulsions between the hydroxyl groups. Thus, NMR spectroscopy has revealed that 1,3 interactions sometimes cause distortion of the zigzag conformation and force the molecules to assume other forms, such as the *sickle* form shown below. In the example illustrated, the sickle conformation is derived from the planar zigzag form by rotation about C-2–C-3 (to alleviate the 1,3-*syn* interaction between the hydroxyl groups on C-2 and C-4).

3

Nomenclature

The currently accepted nomenclature of carbohydrates is the outcome of a continuous international collaborative process aimed at unifying, simplifying, and rationalizing the names of carbohydrates and their derivatives. From the outset, the successive groups of chemists responsible for nomenclature have attempted to (a) follow the general principles used in organic chemistry nomenclature and (b) make as few changes as possible in the terminology in use at the time. Thus, commas, hyphens, and numerals are used in accordance with standard procedures in organic chemistry (commas between numerals and hyphens between letters, but not syllables, as well as between letters and numerals). Similarly, to convert the name of an acid into that of an ester, an acid chloride, an amide, or nitrile, etc., the rules of the International Union of Pure and Applied Chemistry (IUPAC) are followed. To illustrate the second point, namely that changes in terminology are to be avoided whenever possible, it suffices to mention that the trivial names of the common sugars, including aldoses having up to six carbon atoms and ketohexoses, have all been accepted.

In this chapter, the nomenclature of free monosaccharides will be presented first, followed by that of some saccharide derivatives. Finally, the nomenclature of saccharide oxidation and reduction products will be discussed.

I. FREE MONOSACCHARIDES

A. Numbering of a Monosaccharide

The numbering of an acyclic monosaccharide molecule starts at the carbon of, or closest to, the carbonyl group. Accordingly, the carbonyl group is C-1 in an aldose and C-2 in a 2-ketose (counting from the first carbon atom of the chain nearest to the carbonyl group). This makes C-3 the highest numbered carbon atom in an aldotriose, C-4 in an aldotetrose, C-5 in an aldopentose, C-6 in an aldohexose, etc. On cyclization, all of the atoms retain the numbers assigned to the acyclic form.

B. The D and L Configurations

To designate compounds whose configurational relationships to glyceraldehyde have been established, the small capital prefixes D and L are used for pure forms and DL is used for racemic mixtures. These letters are printed in small capital letters or typed doubly underlined, connected by a hyphen before the name of the sugar. If so desired, the D and L notation may be followed by the direction of rotation, enclosed in parentheses, as in (*dextro*), (*levo*), or (*meso*), all italicized in print or typed singly underlined, or by using the signs (+), (−), and (±).

It was mentioned on page 17 that monosaccharides obtainable from D-(+)-glyceraldehyde by ascending reactions belong to the D chiral family and are designated D, and those obtainable from L-(−)-glyceraldehyde belong to the L family and are designated L. The chiral center that determines the D and L notation (and whose configuration is shared by all the members of the same chiral family) is the one farthest from the carbonyl group (usually the penultimate carbon atom in the chain).

C. Aldoses and Chain Configuration

As mentioned earlier, the trivial names of the two aldotrioses (D- and L-glyceraldehyde), and four aldotetroses (D- and L-erythrose and -threose), the eight aldopentoses (D- and L-ribose, -arabinose, -xylose, and -lyxose), and the sixteen aldohexoses (D- and L-allose, -altrose, -glucose, -mannose, -gulose, -idose, -galactose, and -talose) have all been accepted.

The prefixes derived from these names are used to designate the configuration of a chain composed of a group of consecutive (but not necessarily contiguous) chiral carbon atoms to which are attached H, H and OR, H and NHR, or other groups. To name such a group, the italicized prefix derived from the name of the aldose having the same configuration is preceded by the appropriate D or L configurational symbol and followed by the root denoting the total number of carbon atoms in the chain.

```
   R
   |
H-C-OH
   |
  CH2OH
```

D-*glycero*-Dihydroxyethyl-

```
   R                    R
   |                    |
H-C-OH              HO-C-H
   |                    |
H-C-OH               H-C-OH
   |                    |
  CH2OH                CH2OH
```

D-*erythro*-Trihydroxypropyl- D-*threo*-Trihydroxypropyl-

```
   R            R            R            R
   |            |            |            |
H-C-OH      HO-C-H       H-C-OH      HO-C-H
   |            |            |            |
H-C-OH       H-C-OH     HO-C-H       HO-C-H
   |            |            |            |
H-C-OH       H-C-OH      H-C-OH       H-C-OH
   |            |            |            |
  CH2OH        CH2OH        CH2OH        CH2OH
```

D-*ribo*-Tetrahydroxybutyl D-*arabino*-Tetrahydroxybutyl D-*xylo*–Tetrahydroxybutyl D-*lyxo*-Tetrahydroxybutyl

D. Ketoses

Ketoses in which the carbonyl group contains C-2 are named by prefixing one of the foregoing prefixes denoting the configuration (for example, D-*arabino*) to a root denoting the total number of carbon atoms in the chain (for example, hex), and suffixing the whole with "ulose," to make, for example, D-*arabino*-2-hexulose. As already mentioned, the trivial names of the four D-2-hexuloses (D-fructose, D-psicose, D-sorbose, and D-tagatose) are accepted because of established usage.

Ketoses in which the carbonyl group involves a carbon atom higher than C-2 are named by inserting its positional numeral immediately before the root denoting the total number of carbon atoms, for example, D-*arabino*-3-hexulose.

```
    CH2OH                 CH2OH
      |                     |
     C=O               HO-C-H
      |                     |
 HO-C-H                   C=O
      |                     |
  H-C-OH               H-C-OH
      |                     |
  H-C-OH               H-C-OH
      |                     |
    CH2OH                 CH2OH
```

D-*arabino*-Hexulose (D-fructose)   D-*arabino*-3-Hexulose

E. Higher Monosaccharides

To name an aldose or a ketose having more than four chiral carbon atoms (an aldoheptose or higher aldose, or a keto-octose or higher ketose), two or more prefixes are used. Of these, one prefix indicates the configuration of the four chiral carbon atoms adjacent to the carbonyl group, and the other(s) the configuration of the remaining chiral carbon atoms in consecutive groups of four, ending with a group of four or less. In the name of a saccharide, these prefixes are cited in the reverse order; i.e., the first cited is the last unit (farthest from the carbonyl group), followed by the next one, and ending with the one closest to the carbonyl group. For example, for the compound shown below, the prefixes are D-*glycero*-D-*gluco*. These prefixes are added to a root indicating the total number of carbon atoms in the chain (for example, ''hept'') and the whole is suffixed with ''ose'' for aldoses or ''ulose'' for ketoses, making the name D-*glycero*-D-*gluco*-heptose for the compound shown.

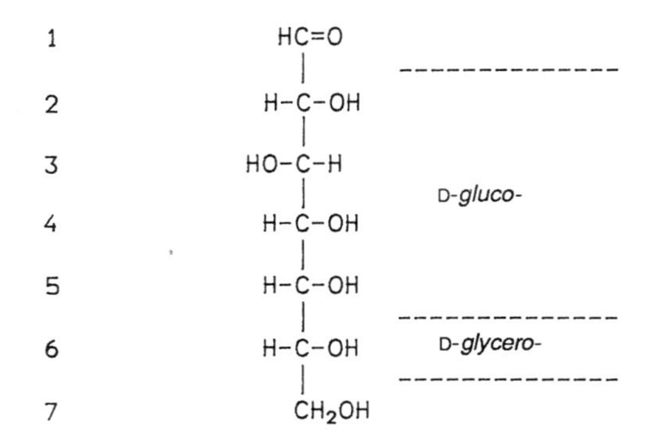

D-*glycero*-D-*gluco*-Heptose

F. Acyclic and Cyclic Forms

The acyclic forms are designated by the italicized prefixes *aldehydo* for aldoses and *keto* for ketoses, inserted immediately before the configurational prefix. For the cyclic forms, the ending "se" of a sugar is replaced by "furanose," "pyranose," or "septanose." If the anomeric configuration is to be designated, the Greek letter α or β precedes and is hyphenated to the D or L symbol; for monosaccharides, the α and β are never used alone without reference to either D or L; for example, α-D-glucofuranose.

In assigning an α or β designation to a glycoside or a nucleoside, using the (*R*), (*S*) system described on page 32 for the free sugars, it should be remembered that in these compounds the OH-1 group of the parent sugar has been replaced by O–R or N–R, respectively, or by C–R in their C analogs; this may alter their priorities relative to other groups and therefore their (*R*) and (*S*) notation. Accordingly, it is recommended that the α, β notation be first assigned to a free sugar of the same configuration and then, when a name is assigned, the ending "ose" is replaced by "oside," to make the word "furanoside" or "pyranoside"—thus, for example, α-D-arabinopyranose and methyl β-D-arabinopyranoside for the compounds depicted.

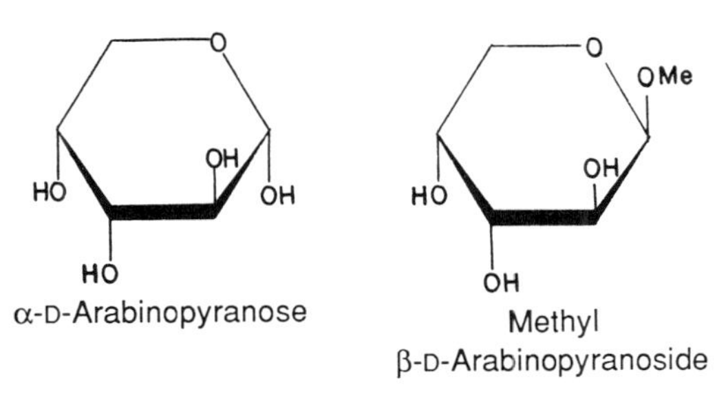

α-D-Arabinopyranose

Methyl β-D-Arabinopyranoside

II. MONOSACCHARIDE DERIVATIVES

A. Substitution at a Carbon Atom

Replacement of a hydroxyl group by hydrogen affords a derivative that is named by hyphenating its positional numeral (locant) to the word "deoxy" followed by the name of the sugar; examples are 6-deoxy-D-galactose and 6-deoxy-L-mannose, which have the accepted names D-fucose and L-rhamnose. It may be noted that the commonly used name 2-deoxy-D-ribose is erroneous and should be replaced by 2-deoxy-D-*erythro*-pentose, because it has only two chiral centers.

The names of the deoxy sugars may be used to name a product in which the hydroxyl group of a sugar has been replaced by an amino group. This is achieved by considering the aminated product as a derivative of the deoxy sugar. Thus, for example, D-glucosamine is called 2-amino-2-deoxy-D-glucose (note that the amino and deoxy are listed alphabetically).

If, instead, one of the hydrogen atoms attached to the carbon skeleton is replaced by a group, the positional numeral followed by an italicized *C* is placed before the name of the group.

```
HC=O
H-C-OH
HO-C-H
HO-C-H
H-C-OH
CH3
```

6-Deoxy-D-galactose (D-fucose)

```
HC=O
CH2
H-C-OH
H-C-OH
CH2OH
```

2-Deoxy-D-*erythro*-pentose (not 2-deoxy-D-ribose)

```
HC=O
H-C-NH2
HO-C-H
H-C-OH
H-C-OH
CH2OH
```

2-Amino-2-deoxy-D-glucose (D-glucosamine)

B. Substitution on Oxygen

1. Ethers and Esters

Replacement of the hydroxyl hydrogen atom by a group is expressed by the locant positional numeral (if needed) followed by an italicized *O* and the name of the group—thus, 2,3,4,6-tetra-*O*-methyl-β-D-glucopyranose and penta-*O*-acetyl-β-D-glucopyranose. In addition, because an ester may

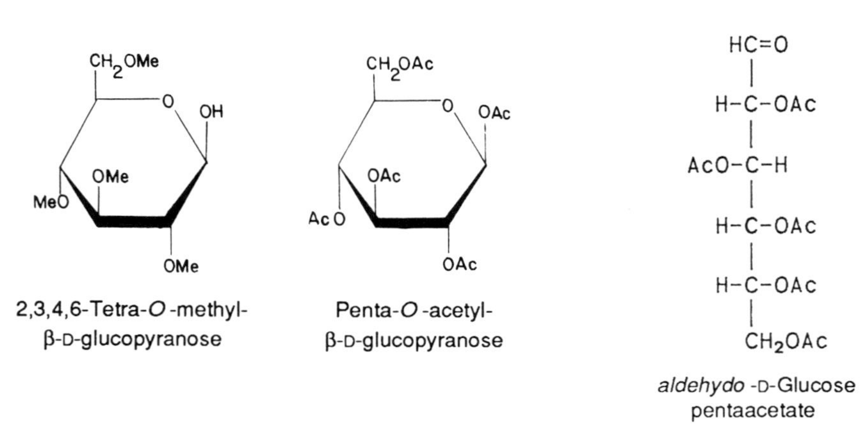

2,3,4,6-Tetra-*O*-methyl-β-D-glucopyranose

Penta-*O*-acetyl-β-D-glucopyranose

aldehydo-D-Glucose pentaacetate

be named by placing, separately after the name of the sugar, the positional numeral, a hyphen, and the ester name, the last-named derivative could also be called β-D-glucopyranose pentaacetate.

2. Acetals and Glycosides

Two types of acetal are possible: (i) cyclic ones, obtained by reaction of the sugar hydroxyl groups with an aldehyde or a ketone, and (ii) those obtained by causing the sugar carbonyl group to react with an alcohol. The latter include both acyclic and cyclic acetals, the latter being termed glycosides.

(i) Cyclic acetals, obtained by causing a pair or pairs of hydroxyl groups of a sugar to react with an aldehyde or a ketone, are named by designating the two positional numerals (separated by commas) to which the group is attached (if more than one group is present, then the pairs of numerals are separated by a colon) followed by an italicized *O*, the IUPAC designation of the group, and lastly the name of the sugar, for example, 1,2:5,6-di-*O*-isopropylidene-α-D-glucofuranose.

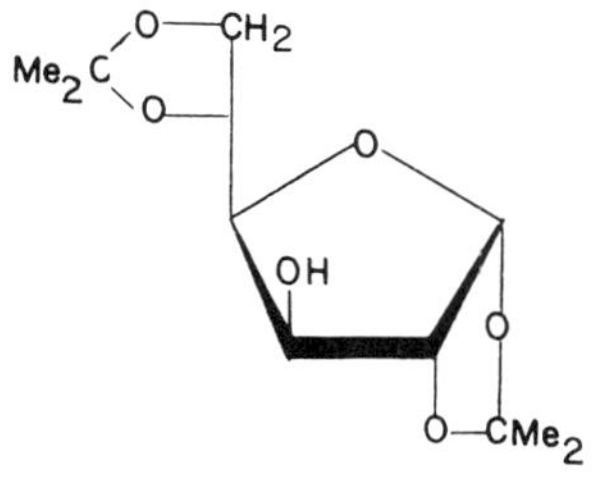

1,2:5,6-Di-*O*-isopropylidene-α-D-glucofuranose

(ii) Acyclic acetals, obtained by nucleophilic attack of an alcohol or a thiol on the carbonyl group of a sugar, are named by inserting the name of the radical between "di" and "acetal" after the name of the sugar—thus,

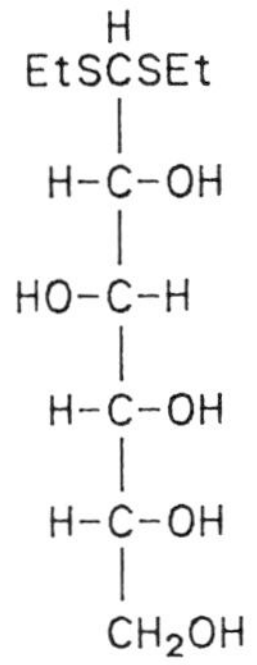

D-Glucose diethyl dithioacetal

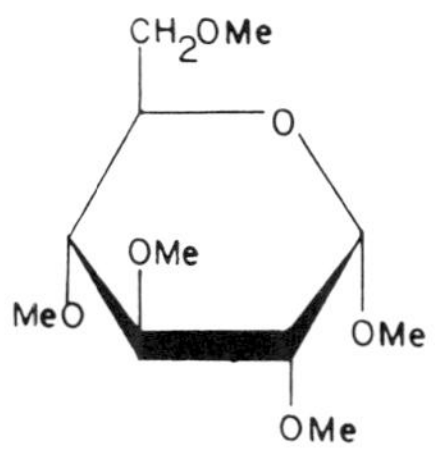

Methyl tetra-*O*-methyl-α-D-glucopyranoside

D-fructose diethyl acetal and D-glucose diethyl dithioacetal. Glycosides are mixed cyclic acetals named by designating the name of the radical (methyl, phenyl, etc.) followed (separately) by the name of the cyclic form of the sugar after replacement of the terminal "e" of the latter by "ide," for example, methyl tetra-*O*-methyl-α-D-glucopyranoside.

III. OXIDATION AND REDUCTION PRODUCTS

A. Dialdoses, Aldosuloses, and Diuloses

Because an aldehyde group is always terminal, no positional numerals are needed for dialdoses, and their names are derived by hyphenating the configurational prefix in italics to the root indicating the chain length, and suffixing the whole with "dialdose." An example is D-*gluco*-hexodialdose.

It may be noted that: (a) Because the dialdose molecule no longer has a top or a bottom, dialdoses that lack an axis of symmetry may be described by either of two configurational prefixes; the one having alphabetic priority is the recommended name. Dialdoses having an axis of symmetry are designated by only one prefix, for example, D-*manno*-hexodialdose. (b) the D and L notation is not normally used for *meso* compounds, such as *ribo*-pentodialdose.

```
   HC=O          HC=O
    |             |
  H-C-OH        HO-C-H           HC=O
    |             |               |
 HO-C-H         HO-C-H          H-C-OH
    |             |               |
  H-C-OH         H-C-OH         H-C-OH
    |             |               |
  H-C-OH         H-C-OH         H-C-OH
    |             |               |
   HC=O          HC=O            HC=O
```

D-*gluco*-Hexodialdose D-*manno*-Hexodialdose *ribo*-Pentodialdose

Sugars possessing an aldehyde group and a keto group, or two keto groups, are named by stating the configurational prefix (in italics), for example, "D-*arabino*," followed by the positional numerals when needed, the radical denoting the number of carbon atoms in the chain, and the suffix denoting the type, namely "osulose" or "diulose"—thus, for example, D-*arabino*-hexosulose and *ribo*-2,6-heptodiulose (it may be noted that the latter, being a meso compound, does not require the D or L notation).

```
   HC=O                 CH2OH
    |                     |
   C=O                   C=O
    |                     |
 HO-C-H                H-C-OH
    |                     |
  H-C-OH               H-C-OH
    |                     |
  H-C-OH               H-C-OH
    |                     |
   CH2OH                 C=O
                          |
                        CH2OH
```

D-*arabino*-Hexos-2-ulose *ribo*-2,6-Heptodiulose

B. Alditols and Aldaric Acids

Alditols are polyhydroxy alcohols obtained by reduction of monosaccharides, and aldaric acids are polyhydroxy dicarboxylic acids obtained by vigorous oxidation of monosaccharides. The former are named by replacing the suffix "ose" with "itol" and the latter by replacing it with "aric acid." Like the dialdoses, described earlier, alditols and aldaric acids lacking an axis of symmetry can be derived from each of two aldoses and may be given one of two names, whereas those having an axis of symmetry can be assigned only one name. Likewise, compounds having a plane of symmetry (meso) do not require the D or L notation.

```
   CH2OH
    |
  H-C-OH                COOH
    |                     |
 HO-C-H                H-C-OH
    |                     |
  H-C-OH              HO-C-H
    |                     |
  H-C-OH               H-C-OH
    |                     |
   CH2OH                 COOH
```

D-Glucitol or L-gulitol Xylaric acid

C. Aldonic Acids

Monocarboxylic acids obtained by oxidation of the aldehydic group to a carboxylic group are named by replacing the ending "ose" in the name of the sugar by "onic acid." For acid derivatives, the words "chloride," "amide," etc. may be added. To derive from these names the names of esters, salts, nitriles, etc., the usual organic chemistry system for these groups is used.

```
                        COOMe                CN
                          |                   |
     COOH               H-C-OH              H-C-OH
      |                   |                   |
    HO-C-H              HO-C-H              HO-C-H
      |                   |                   |
     H-C-OH             HO-C-H              H-C-OH
      |                   |                   |
     H-C-OH              H-C-OH             H-C-OH
      |                   |                   |
     CH2OH              CH2OH               CH2OH
```

D-Arabinonic acid — Methyl D-galactonate — D-Glucononitrile

D. Glycuronic and Glyculosonic Acids

Oxidation of the primary hydroxyl group in an aldose affords the corresponding uronic acid, and oxidation of one such group in a ketose affords a ulosonic acid. The first are named by substituting the ending "ose" in the aldose name by "uronic acid." Examples are D-glucuronic acid and methyl D-mannopyranuronate.

To name the keto acids, the ending "ulosonic acid" is preceded by the configurational prefix and the root designating the total number of carbon

```
     HC=O               HC=O
      |                   |
     H-C-OH             HO-C-H               COOH
      |                   |                   |
    HO-C-H              HO-C-H              H-C-OH
      |                   |                   |
     H-C-OH              H-C-OH              C=O
      |                   |                   |
     H-C-OH              H-C-OH             H-C-OH
      |                   |                   |
     COOH               COOMe               CH2OH
```

D-Glucuronic acid — Methyl D-mannuronate — D-*erythro*-3-Pentulosonic acid

atoms. If the keto group is in a position other than 2, its positional numeral is inserted before the last mentioned. Examples are D-*arabino*-2-hexulosonic acid and D-*erythro*-3-pentulosonic acid.

PROBLEM

1. Give the correct names for the following monosaccharide derivatives.

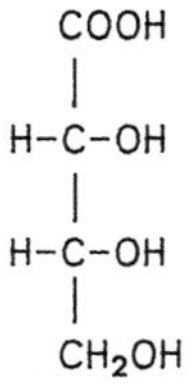

(1)

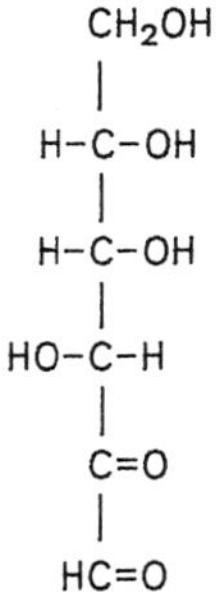

(2)

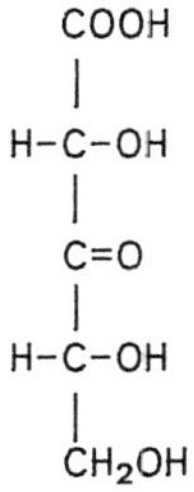

(3)

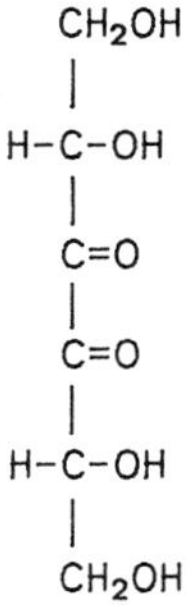

(4)

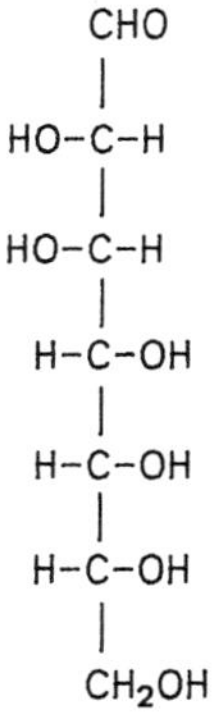

(5)

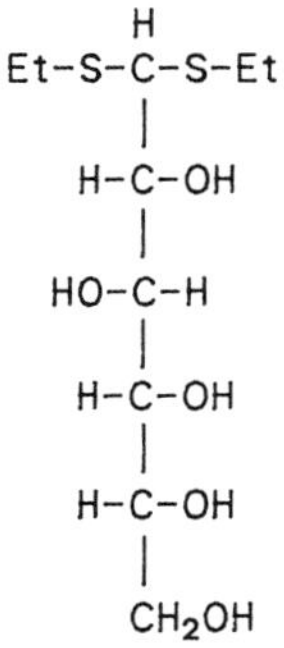

(6)

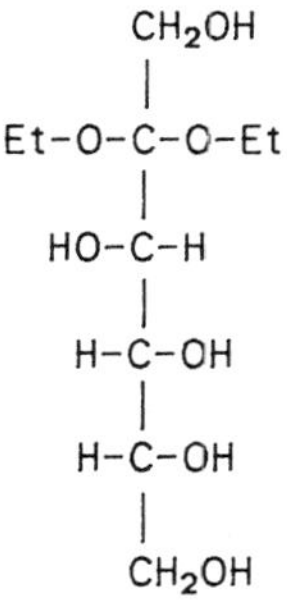

(7)

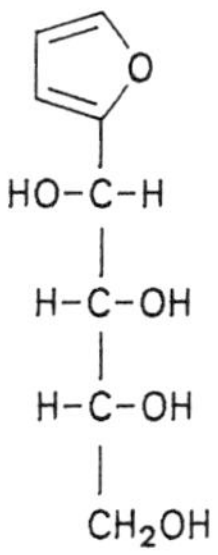

(8)

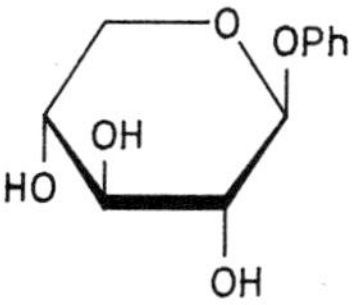

(9)

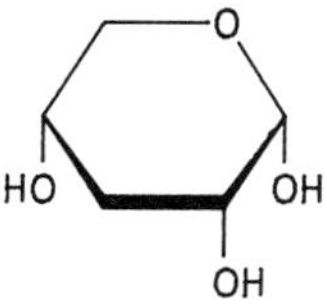

(10)

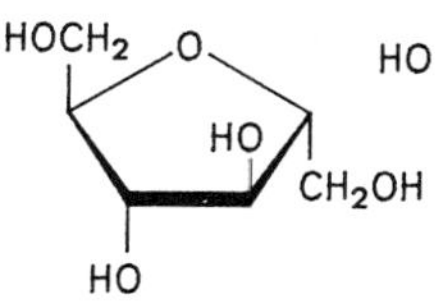

(11)

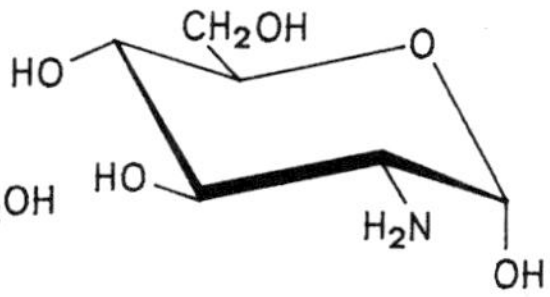

(12)

4

Physical Properties Used in Structure Elucidation

Spectroscopic and polarimetric methods have been used extensively to elucidate the structure of saccharides and their derivatives. The order in which the different spectroscopic methods are treated in this chapter parallels the amount of energy required to reach their excited states. First discussed will be nuclear magnetic resonance (NMR) spectroscopy, which detects low-energy absorptions in the radio-frequency range and includes one-dimensional ^{1}H-, ^{13}C-, and ^{15}N-NMR spectroscopy as well as two-dimensional NMR techniques. This is followed by molecular and electronic spectroscopy, which deals with absorptions in the infrared, visible, and ultraviolet regions. Mass spectrometry is discussed next. Here, the energy provided to ionize molecules often exceeds the dissociation energy of covalent bonds and causes fragmentation to occur. Finally, polarimetric methods are reviewed, which use polarimeters, optical rotatory dispersion (ORD), and circular dichroism (CD) instruments to measure changes in the ellipticity or absorbance of polarized light as it passes through chiral molecules. Spectroscopic methods that involve reversible promotions from ground states to excited states, such as NMR spectroscopy, are nondestructive, whereas those involving irreversible breaking of bonds, as in mass spectrometry, are destructive.

I. NUCLEAR MAGNETIC RESONANCE SPECTROSCOPY

When subjected to an external magnetic field, nuclei having a spin of $\frac{1}{2}$, 1, or $1\frac{1}{2}$ are said to be in a lower energy level, designated the *ground state*, when oriented parallel to the field. When these nuclei absorb energy in the form of appropriate radio frequencies, they flip around, assuming an antiparallel orientation, and in so doing are promoted to a higher energy level, the *excited state*. The amount of energy needed to cause the flipping of nuclei depends on the frequency of the radio signal, the strength of the magnetic field, and the type of nucleus and its chemical environment. Most NMR spectrometers used for structure elucidation measure the resonance of nuclei having a spin of $\frac{1}{2}$, such as ^{1}H, ^{13}C, ^{15}N, ^{19}F, and ^{31}P, which give well-resolved spectra, unlike the ones produced by nuclei having higher spins. Some commercial instruments measure the resonance frequency at constant magnetic field generated by permanent magnets, whereas others measure the resonance magnetic field at constant frequency using variable electromagnets. They operate at a frequency of 90, 200, 400, or 500 MHz and often require superconducting solenoids.

A. ^{1}H-Nuclear Magnetic Resonance Spectroscopy

^{1}H-NMR is by far the most widely used type of NMR spectroscopy. This is why most laboratories dedicate one or more instruments to the study of the ^{1}H nucleus.

The frequency at which a given proton resonates (flips to the antiparallel orientation), relative to a standard, is designated the *chemical shift*. The extent of this shift depends on the environment of the proton, which in turn is determined by the structure of the molecule. The electrons present in a group adjacent to a resonating proton affect the frequency at which the latter resonates by shielding or deshielding it from the external magnetic field, which increases or decreases the energy needed to flip it around, i.e., the resonance frequency. Shielding occurs when the circular motion of adjacent electrons, induced by the external magnetic field, creates a small magnetic field that opposes the outer field. This phenomenon is exhibited by all magnetically isotropic groups. Deshielding occurs when electrons are pulled away from a resonating proton by electronegative groups or atoms, rendering the proton more susceptible to the outer magnetic field.

The chemical shift of a given proton changes with the strength of the magnetic field and is expressed relative to a standard, usually tetramethylsilane (Me_4Si), in dimensionless values expressed in parts per million (ppm) as either δ or τ ($\tau = 10 - \delta$) as follows.

$$\delta = \frac{10^6 \times \text{frequency difference between standard and reference (Hz)}}{\text{frequency of oscillator (Hz)}}$$

1. Spin–Spin Coupling and Coupling Constants

The spin of an adjacent proton affects the resonance frequency of a resonating proton. This occurs because a spinning proton generates a magnetic field, which can enhance the outer magnetic field if parallel to it or diminish it if antiparallel to it. In addition, two protons having opposing spins may cancel each other's effect on the field. The number of ways in which protons can be aligned with or against the outer magnetic field, as well as the ratio of each combination with respect to the other ones, is predicted by the laws of probability. The small changes in the local magnetic field caused by the spin of adjacent protons result in splitting of the resonance signals, a phenomenon called *spin–spin coupling*. The pattern of splitting is of value in structure elucidation, as it provides a count of the protons adjacent to the resonating proton. Thus, a proton (a methyne group) will produce a doublet, two protons (a methylene group) a triplet, three protons (a methyl group) a quartet, etc. The spacing, measured in hertz, between the branches of a split signal is known as the *coupling constant* and is designated by the letter J subscripted by the numerals of the two interacting protons, for example, $J_{1,2}$. Unlike chemical shifts, which depend on the frequency of the oscillator, coupling constants are unaffected by the oscillator frequency. The coupling constant is a function of the dihedral angle between the two adjacent protons, being maximal at 0° and 180° angles and minimal at a 90° angle. The Karplus equation, which governs the relationship between coupling constants and dihedral angles, has been used extensively in carbohydrate chemistry to determine the configuration and conformation of saccharide derivatives.

Elucidation of the structure of a saccharide by ^{1}H-NMR usually starts with a study of the number and integration of the signals to determine whether they account for all of the protons in the molecular formula and whether their ratios agree with the expected structure. This is followed by a study of the splitting pattern of the different signals, to determine the number of protons adjacent to each resonating proton or group of protons (one adjacent proton for a doublet, two for a triplet, three for a quartet, etc.). The signals in the ^{1}H-NMR spectra of most monosaccharide derivatives can readily be assigned to the appropriate protons by making use of the observed chemical shifts and splitting patterns (see Table I). To confirm these assignments, decoupling experiments are carried out. These consist of consecutively irradiating (saturating) the sample with the resonance frequencies of the individual signals and observing the collapse in

TABLE I
NMR Characteristics of Some D-Hexopyranoses

Hexopyranose	H-1 ($J_{1,2}$)	H-4
α-D-Glucopyranose	Doublet (4 Hz)	Doublet of doublet
β-D-Glucopyranose	Doublet (8 Hz)	Doublet of doublet
α-D-Mannopyranose	Doublet (3 Hz)	Doublet of doublet
β-D-Mannopyranose	Doublet (2 Hz)	Doublet of doublet
α-D-Galactopyranose	Doublet (4 Hz)	Multiplet
β-D-Galactopyranose	Doublet (8 Hz)	Multiplet

the splitting pattern of the signals of the adjacent protons. Thus, saturation of the readily identifiable anomeric-proton signal (usually a low-field doublet) will cause the collapse of the H-2 multiplet to a simpler splitting pattern (now split by H-3 only). Irradiation at the H-2 frequency will then identify the H-3 signal, which in turn is irradiated to identify H-4, and so on, until all of the protons on consecutive carbon atoms have been identified.

The configuration and conformation of a monosaccharide can then be determined by examining the coupling constant (J) of each of the signals and estimating from the Karplus equation (see page 36) the dihedral angle between adjacent protons, starting with the anomeric position (usually a low-field doublet). Finally, the proximity of one proton to another in a given conformation may be determined from the nuclear Overhauser effect (NOE), which results from the transfer of magnetization from one nucleus to another. The NOE causes the intensity of the resonance of a nucleus to change if the z component of magnetization of a nearby proton is perturbed. Usually the magnetization is perturbed selectively by irradiation with a weak radio-frequency field, and the change in the intensities of the other resonances in the spectrum is monitored. Hydrogen bonding can readily be detected by the deshielding observed in the deuteratable OH and NH signals. To differentiate between intra- and intermolecular hydrogen bonding, the spectra are measured in solvents of different polarity and at different dilution.

Figure 1 shows a high-resolution ^{1}H-NMR spectrum of β-D-glucopyranose before and after decoupling of the OH protons. It may be noted that the anomeric proton of the α form is present in traces at δ 4.9.

B. ^{13}C-Nuclear Magnetic Resonance Spectroscopy

With the increased availability of multinuclear NMR spectrometers, ^{13}C-NMR spectra are now measured routinely to confirm structures, con-

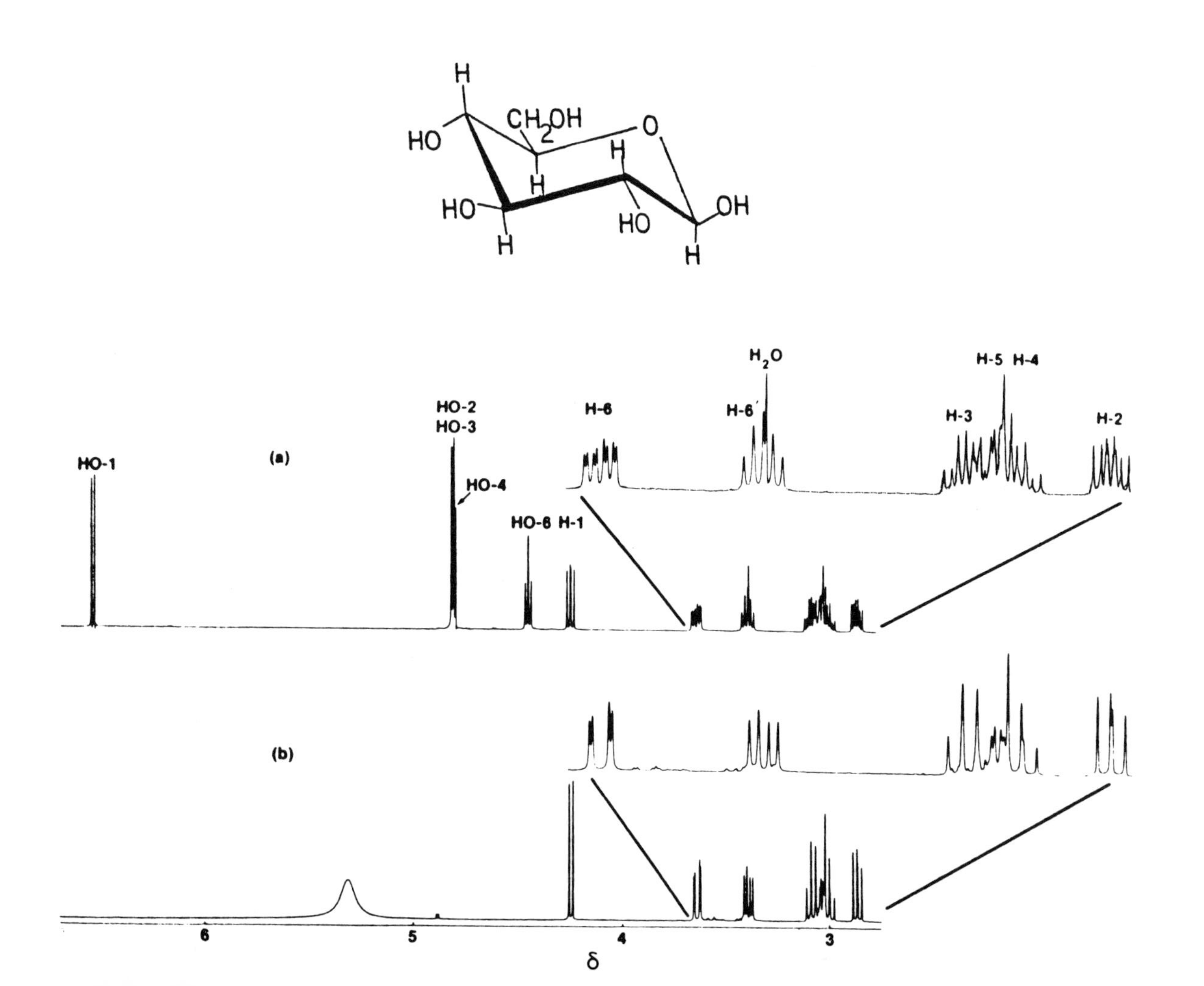

H
HO
CH2OH
O
H
H
HO
HO
OH
H
H
HO-1
(a)
HO-2
HO-3
HO-4
HO-6 H-1
H-6
H-6´
H2O
H-3
H-5 H-4
H-2
(b)
6
5
4
3
δ

figurations, and conformations of saccharides. An advantage of this type of NMR spectroscopy over ^{1}H-NMR is that the signals of the different carbon atoms seldom overlap and each carbon atom in the molecule can be accounted for. This is because there are fewer signals (carbohydrates have fewer carbon atoms than protons) spread over a wider range of frequency. However, since the natural abundance of ^{13}C is quite small (the ratio of ^{13}C to ^{12}C is 1.11 : 98.89), much larger samples are needed than are required for ^{1}H-nuclear magnetic resonance spectra. To enhance the quality of the spectra, the signals generated by a pulsating oscillator are analyzed by use of Fourier transforms (FTs) and time-averaged.

Because the magnetic field required for ^{13}C-NMR spectroscopy is much stronger than that needed for ^{1}H-NMR, signals in the latter couple, but do not interfere, with those in the former. The spinning protons directly attached to the resonating carbon atoms produce coupling constants much larger than those in ^{1}H-NMR spectroscopy. As a rule, two ^{13}C-NMR spectra are measured for each sample, one recorded with proton decoupling and one without.

There are several similarities between ^{1}H- and ^{13}C-NMR spectra; thus, if a certain carbon atom resonates at a higher field than another, it is probable that the hydrogen atoms attached to the first carbon atom will resonate at a higher field than those attached to the second carbon atom. For example, the ^{1}H and ^{13}C signals of a methyl group appear at a higher field than those of a methylene or a methyne group. This occurs because the chemical shifts of carbon and hydrogen are affected in the same way by shielding and deshielding. It should be noted, however, that some groups that appear in ^{13}C spectra may not be seen in ^{1}H-NMR spectra, and vice versa, which renders ^{13}C- and ^{1}H-NMR spectra complementary. For example, carbonyl groups, which appear at low field in the ^{13}C-NMR spectra, are not visible in ^{1}H-NMR spectra, and OH and NH groups, detectable in the latter spectra, do not appear in the former.

An area where ^{13}C-NMR spectroscopy has yielded better results than ^{1}H-NMR is in the determination of the α or β configuration of furanosides. Because the anomeric proton–proton coupling constants of the two anomers are quite close, a conclusive assignment is often based on the ^{13}C-NMR spectrum of the 2,3,O-isopropylidene derivative. The chemical shift of the *gem*-dimethyl carbon atoms in the rigid bicyclic isopropylidene furanoside were found to vary little. They range from 24.9 to 26.3 ppm for the α anomer and 25.5 to 27.5 ppm for the β anomer. Figure 2 shows the ^{13}C-NMR spectrum of methyl β-D-xylopyranoside in D_2O.

Fig. 1. ^{1}H-NMR spectra of β-D-glucose in Me_2SO-d_6 at 400 MHz. (a) OH protons coupled; (b) OH protons decoupled by CF_3CO_2H. (Original provided by Dr. B. Coxon.)

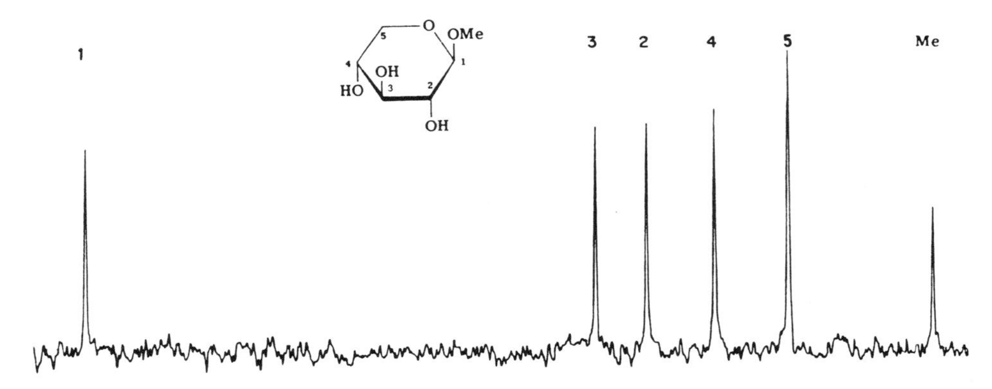

Fig. 2. ^{13}C-NMR spectrum of methyl β-D-xylopyranoside in D_2O. (Original provided by Dr. B. Coxon.)

C. ^{15}N-Nuclear Magnetic Resonance Spectroscopy

Natural-abundance ^{15}N-NMR spectroscopy has found fewer applications in the study of carbohydrates than the two types of NMR spectroscopy discussed earlier. Except for amino sugars and such nitrogen derivatives as oximes and hydrazones, nitrogen is not often found in carbohydrates. However, because ^{15}N-NMR spectra provide valuable information on the structure of nitrogen-containing saccharides, this type of spectroscopy is becoming more widely used. As in ^{13}C-NMR spectroscopy, the protons attached to nitrogen produce large couplings in the ^{15}N signals, which may be decoupled if desired. Figure 3 shows the ^{15}N-NMR spectrum of a bis(phenylhydrazone) of L-ascorbic acid.

D. Two-Dimensional Nuclear Magnetic Resonance Spectroscopy

The technique of two-dimensional NMR spectroscopy has evolved into a valuable tool for elucidation of the structure of monosaccharides, oligosacharides, and polysaccharides. In one-dimensional NMR spectroscopy, the abundance of the lines observed often makes it difficult to assign each resonance signal to its corresponding nucleus. The power of 2-D NMR lies in its ability to resolve overlapping spectral lines, to enhance sensitivity, and to provide information not available by 1-D methods. In addition, 2-D NMR enables the measurement of internuclear distances and the coupling constants J in molecules that are too complex for study by the 1-D approach.

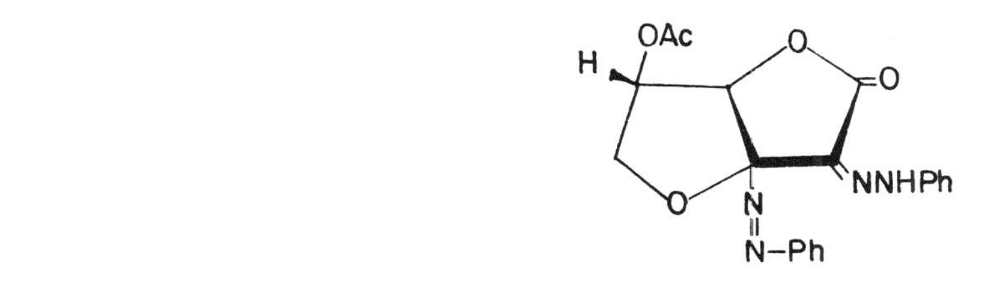

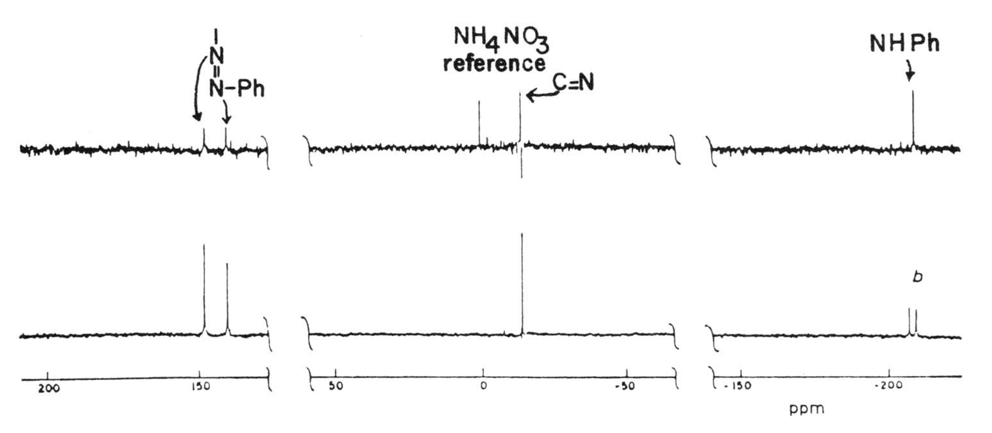

Fig. 3. Natural-abundance ^{15}N-NMR spectra at 40.5 MHz of an acetylated oxidation product of L-ascorbic acid bis(phenylhydrazone) in chloroform-*d*. (Upper spectrum proton-decoupled; lower spectrum proton-coupled. (Original provided by Dr. B. Coxon.)

Two-dimensional NMR spectroscopy is based on the separation of chemical shift and coupling effects along two frequency axes to simplify the analysis of overlapping spectral lines. The two frequency dimensions of 2-D NMR originate from the two time intervals, t_1 and t_2, during which the nuclei can be subjected to two different sets of conditions. The amplitude of the signals detected during time t_2 is a function of what happened to the nuclei during the evolution period t_1. If the experiment is repeated for a large number of incremented t_1 durations, varying from zero to several hundred milliseconds, a set of spectra is obtained with the amplitude of the resonances modulated with the frequencies that existed during period t_1. A Fourier transformation with respect to t_1 defines the modulation frequencies and results in a 2-D spectrum.

It is also possible to transfer magnetization from one nucleus to another by the nuclear Overhauser effect or through the scalar coupling (*J* coupling) mechanism. Interactions between the magnetic dipole moments of two adjacent nuclei cause the intensity of the resonance of the first nucleus to change if the z component of the magnetization of the adjacent nucleus is perturbed. This nuclear Overhauser effect has been used in one-dimensional NMR spectroscopy, as already mentioned. The cross-relaxation correlation method called nuclear Overhauser effect spectroscopy (NOESY) relies on the same through-space dipolar relaxation. The advantages of NOESY are that (a) it reduces the number of measurements needed, since all the distances are measured simultaneously and not successively as in 1-D experiments, and (b) the spectral overlaps that make selective perturbation impossible in complex molecules are significantly diminished.

Another way to transfer magnetization from one nucleus to another is through the scalar coupling mechanisms, which rely on through-bond vicinal couplings between protons. For example, in coupling correlation spectroscopy (COSY), the pulse sequence starts when a $\pi/2$ pulse turns the magnetization vector from its equilibrium position along the z axis into the x–y plane. The transverse magnetization evolves during the delay t_1, at the end of which a second $\pi/2$ pulse mixes the magnetizations and the chemical shifts. A free induction decay (FID) signal is acquired during the time t_2. The resonances detected by the first Fourier transformation of the FIDs in the t_2 dimension are modulated as a function of t_1 by all frequencies (due both to chemical shifts and coupling constants) that are generated by the first pulse. For transitions that generate resonances in the same multiplet, the chemical shifts that undergo mixing are identical, so that the multiplets in the 2D spectrum have the same chemical shift in both dimensions. Thus, one part of the 2-D COSY spectrum consists of a set of resonances along the diagonal that correspond to the normal one-dimensional spectrum. Other transitions of spin-coupled nuclei are progressively connected through a common energy level, and mixing of their magnetizations produces resonances in the 2-D COSY spectrum that have different chemical shifts in each dimension. These resonances therefore occupy off-diagonal positions and are known as cross peaks. The observation of cross peaks signifies spin couplings between the related multiplets on the diagonal. Figure 4 shows the ^1H–^1H– 2-D COSY spectrum of 2,6-dideoxy-β-D-*ribo*-hexose (β-D-digitoxopyranose).

By using 2-D ^1H-NMR experiments obtained by homonuclear chemical shift correlation (COSY), a complete assignment of the complicated proton spectrum is possible. This in turn can lead to ^{13}C assignment by means of ^1H–^{13}C heteronuclear correlation maps. It is now possible to determine the structure and conformation of molecules having a molecular weight of about 15,000 in a nondestructive manner with such techniques. [For a review on 2-D NMR spectroscopy, see A. Bax and L. Lerner, *Science,* **232,** 960 (1986).]

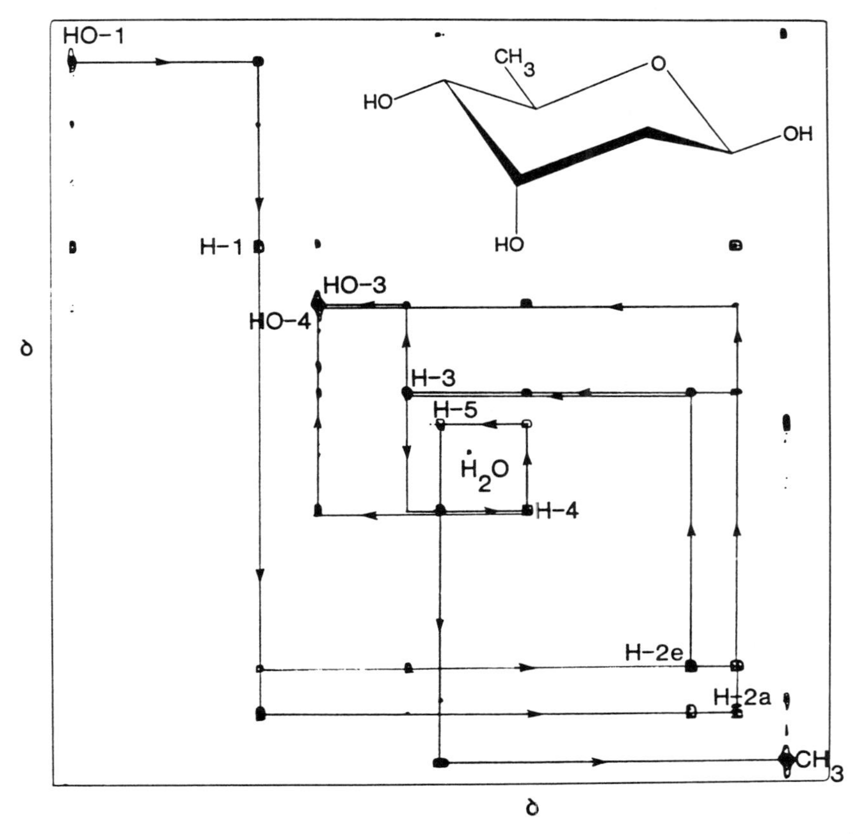

Fig. 4. Contour plot of the 2-D COSY ^{1}H-NMR spectrum of β-D-digitoxopyranose in DMSO-d_5 at 400 MHz, with the spin-coupling connectivity pathway indicated. (Original provided by Dr. B. Coxon.)

II. MOLECULAR SPECTROSCOPY

Transitions between different vibrational and rotational levels in organic molecules require energy which corresponds to frequencies in the near-infrared and the infrared for the former and the infrared and microwave for the latter. Vibrational modes include stretching, bending, twisting, scissoring, breathing, etc. For structure elucidation, stretching vibrations are useful because they are characteristic of bond types and always appear at the same wavelength. For example, the stretching of an O–H group always occurs between 2.7 and 3. 4 μm. When hydrogen is replaced by a heavier element, the stretching wavelength is shifted to higher values (lower energies); thus the O–C and C–C stretching appears at 6.7–15 μm.

On the other hand, double and triple bonds, which are stronger bonds than single C–C bonds and require more energy to stretch, absorb at 5.2–6.7 μm for the former and 4.2–5.0 μm for the latter.

Most commercial infrared spectrophotometers are double-beam recording instruments that measure the percent absorption versus wavelength in micrometers; or wave number in reciprocal centimeters (cm^{-1}), in the range of 2.5 μm (4000 cm^{-1}) to 15 or 25 μm (665 or 400 cm^{-1}).

For carbohydrates and their derivatives, samples are usually mixed with KBr and pressed into transparent pellets. Alternatively, they may be suspended in paraffin (Nujol) to form a mull, which is spread between two salt plates, or dissolved in an organic solvent and introduced into a cell made of a sodium or potassium halide. The spectra produced by the last two methods will show the absorptions of both the sample and the liquid used to disperse it.

There are two distinct parts in an IR spectrum. The first, and most useful for structure elucidation, ranges from 2.5 to 6.7 μm and comprises stretching vibrations characteristic of the groups present in the molecule. The most readily recognizable of these are caused by O–H, C–H, C═O, C═C, C═N, and C═C groups. The second part, called *the fingerprint region,* ranges from 6.7 to 15 μm and comprises stretching and other vibrations, which usually crowd the spectrum with absorption bands and can be used to identify a compound.

Infrared spectroscopy has been used extensively to elucidate aspects of the structure of monosaccharides. For example, in Chapter 2 we saw that the absence of a carbonyl absorption from the IR spectrum of D-glucose was used to show that the acyclic (aldehyde) form of this sugar is not present to any appreciable extent, because it equilibrates to give different cyclic forms. Infrared spectra can differentiate between 1,4- and 1,5-lactones and detect esters and amides by their carbonyl absorptions. Because of the large number of OH groups (many of which are hydrogen-bonded), the OH stretching vibrations usually appear as broad bands. The nitrile and acetylene stretching vibrations are, of course, useful diagnostic bands for derivatives containing such groups.

III. ELECTRONIC SPECTROSCOPY

When organic molecules absorb energy in the ultraviolet and visible regions, electrons in sigma (σ), pi (π), or nonbonding (n) orbitals undergo promotion from the ground state to higher-energy antibonding levels. The antibonding orbitals related to the first two bonds (σ and π) are designated by asterisks (*); of course, there are no antibonding orbitals associated with nonbonding electrons.

In general, the bands associated with transitions from nonbonding states to antibonding π orbitals ($n \rightarrow \pi^*$) are weaker than those arising from transitions of π electrons to their antibonding orbitals ($\pi \rightarrow \pi^*$) and therefore appear at longer wavelengths than the latter. They also undergo a blue (hypsochromic) shift on changing from a nonpolar to a hydroxylic solvent, because the nonbonding lone pair of electrons is subject to hydrogen-bond formation with the solvent molecules and requires additional energy for the promotion to π^* orbitals.

The $n \rightarrow \pi^*$ transitions are only slightly affected by conjugation, which lowers the transition energy of $\pi \rightarrow \pi^*$ bands by raising the donor orbitals and lowering the acceptor orbitals. This causes a marked bathochromic shift (shift toward longer wavelengths) in the latter and explains why the $\pi \rightarrow \pi^*$ bands can overlap the $n \rightarrow \pi^*$ bands when a conjugated system is extended.

In nonpolar solvents, the $n \rightarrow \pi^*$ and $\pi \rightarrow \pi^*$ bands may both show vibrational fine structure, which is retained in hydroxylic solvents by the latter and lost by the former because of hydrogen bonding with the heteroatom. This fact can be used to assign the correct transitions to absorption bands.

The transitions between nonbonding and antibonding σ electrons ($n \rightarrow \sigma^*$) in a heteroatom usually give rise to strong bands in the 200-nm region, often below the measured spectral range. Nitrogen absorptions occur at somewhat longer wavelengths than oxygen absorptions, because the pair of nonbonded electrons in oxygen occupies a lower energy level than those in nitrogen so that the transitions of the σ^* level requires more energy. In the case of sulfur, the electrons are more weakly bound and excitation of the nonbonding electrons requires less energy. Accordingly, the $n \rightarrow \sigma^*$ bands of saturated sulfur compounds appear at a longer wavelength (200 nm), well within the range of ordinary spectrophotometers. On the other hand, bands arising from transitions of σ electrons to their antibonding orbitals ($\sigma \rightarrow \sigma^*$) occur in the vacuum ultraviolet and can be detected only with special spectrophotometers.

An ultraviolet spectrum is a continuous record of absorbance A versus wavelength λ. Absorbance is defined as A-log I_0/I, where I_0 is the intensity of the incident light and I the intensity of the transmitted light. The Beer–Lambert law states that $A = \varepsilon cl$, where ε is the molar extinction coefficient, c the molar concentration, and l the path length in centimeters. Percent transmission ($I/I_0 \times 100$) versus wavelength is frequently recorded for compounds of unknown molecular weight.

Most UV spectrophotometers record absorbance or transmission curves between 190 and 1000 nm and are capable of repeating the spectrum or a portion thereof at fixed intervals, permitting the monitoring of reaction kinetics. Because of the experimental difficulty involved and the need for special apparatus, carbohydrate derivatives have seldom been studied in the vacuum ultraviolet or in the near-infrared region, both of which merit greater attention.

The lower-wavelength cutoffs for solvents commonly used for carbohydrates and their derivatives are water, 191 nm; hexane and cyclohexane, 195 nm; ethanol, 204 nm; and chloroform, 237 nm, for 1-cm cells.

Free sugars are not normally found in the acyclic (aldehydo or keto) forms to any appreciable extent, but exist mainly in cyclic hemiacetal forms, which do not have π electrons. Accordingly, the longest-wavelength absorptions they exhibit arise from the promotion of nonbonding electrons (from the lone pair on an oxygen atom) to antibonding σ orbitals ($n \rightarrow \sigma^*$). These absorptions usually appear at 200 nm, and are so close to the limit of most instruments that only their tail end is measurable. For this reason, UV spectroscopy is of little use in structural studies of free sugars. On the other hand, many sugar derivatives possess strong chromophores, such as the C=C, C=N, N=N, C=O, C=S, and S=O groups, which exhibit $\pi \rightarrow \pi^*$ and $n \rightarrow \pi^*$ absorptions that are well within the UV–visible range of commercial spectrophotomers. Figure 5

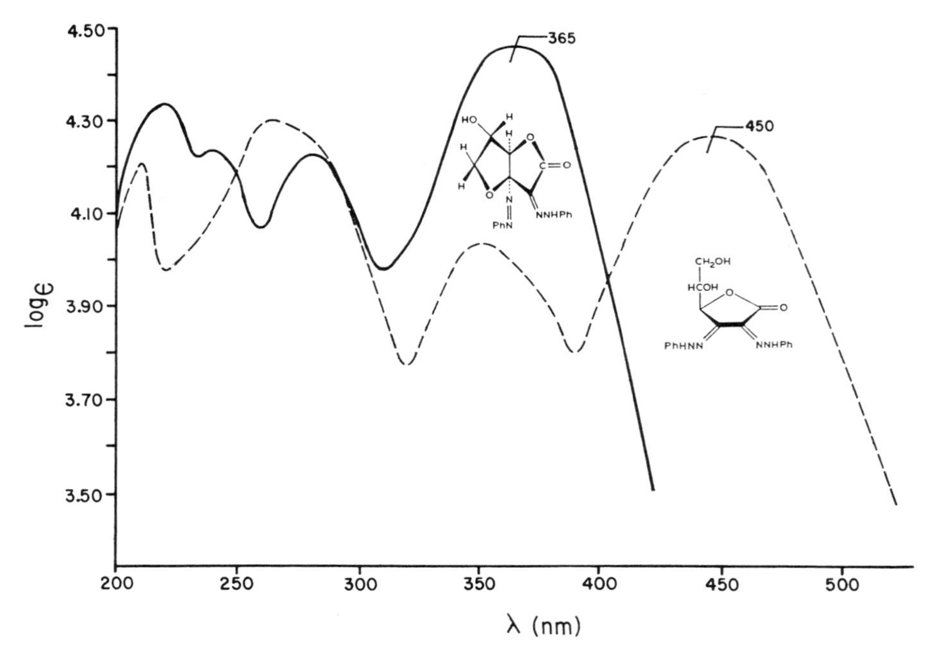

Fig. 5. UV spectrum of two hydrazine derivatives, showing the effect of conjugation on the absorption maxima.

depicts the UV spectrum of two hydrazine derivatives, showing the effect of conjugation on the absorption maxima.

IV. MASS SPECTROMETRY

During the 40 years following its discovery by Aston, mass spectrometry (MS) was used mainly to determine the atomic weights of isotopes. Later, it was discovered that with some modifications, mass spectrometers could be used to measure accurately the molecular weights of organic molecules. What was needed for such measurements was a device to ionize the organic compound, so that the resulting charged particles could be deflected by the magnetic and electric fields of the instrument.

The first method used to ionize organic compounds was electron impact (EI). Here, the sample (about 0.1 mg) is thermally volatilized by heating to 130–200°C under reduced pressure and then bombarded with a beam of electrons. This causes the extraction of one of the loosely bound electrons, usually one of the lone pair of electrons attached to an oxygen atom in a carbohydrate molecule, which generates a positive ion. The resulting *molecular ion* then undergoes a series of rearrangements that cause bond cleavage and the formation of a number of neutral and positively charged fragments. Only the latter are deflected by the magnetic and electric fields of the instrument and are detected. The extent of deflection of the resulting fragment will depend on the ratio of the mass to the charge (m/z).

Milder ionization techniques have been developed to minimize the complete degradation of the molecular ion and its disappearance from the spectrum. For example, in *chemical ionization* (CI) mass spectrometry, the volatilized sample is brought in contact with an ionized gas such as methane, isobutane, or ammonia, which produces low-energy proton-donating ions such as H^+, and $C_2H_5^+$, $C_9H_{11}^+$, and NH_4^+ that can react with the organic molecule, yielding not only the molecular ion M^+ but also ions of higher molecular weight, such as MH^+, $(M + CH_3)^+$, and $(M + C_2H_5)^+$. Despite the production of so many ionic species, CI-MS constitutes one of the best methods of studying such delicate molecules as free monosaccharides and their derivatives, whose molecular ions cannot be detected by EI-MS. Of course, it is possible by using appropriate detectors to perform both positive and negative chemical ionization mass spectrometry. However, the former is much more commonly used; negative CI-MS is advantageous when such strongly electronegative atoms as halogens are present in the molecule.

Other ionization techniques useful in detecting molecular ions are *field ionization* (FI) and *field desorption* (FD). In field ionization, the sample is inserted in a metal compartment having a microscopically narrow edge. In field desorption, the sample is subjected to an intense electric field, which ionizes it and permits the molecules to volatilize with minimal degradation. An improvement on these techniques is *fast atom bombardment* (FAB) mass spectroscopy, in which a liquid matrix is used to ionize large polar molecules without prior chemical derivatization. Molecular weights of about 2500 can be determined accurately in the form of base peaks (M + 1). By using tandem mass spectrometry (MS/MS), it is possible to determine the molecular weight by FAB in the first mass spectrometer and study the daughter ions in the second one. [For reviews on the use of FAB in structure elucidation, see K. Biemann, *Anal. Chem.* **58,** 1288 (1986); A. Dell, *Adv. Carbohydr. Chem. Biochem.* **45,** 19 (1987)].

Because the energy needed to ionize an organic molecule often exceeds the energy necessary to cleave its weaker bonds, degradation invariably occurs in a mass spectrometer. The structure of the molecule will dictate where a shift of electrons is likely to occur in the molecular ion that may cause a bond to break. The stability of the resulting fragments will dictate the magnitude of their signals and thus determine the fragmentation pattern. Mass spectra can be used to identify molecules by comparing their fragmentation patterns with those of known compounds, or to study their structures by verifying the compatibility of their fragmentation patterns with the assigned structures. Mass spectrometry is, however, insensitive to stereochemical changes.

Two types of mass spectrometer are available commercially, one designed to detect and identify organic compounds and the other to permit study of the structure of such compounds. In the first type, the intensity and mass of a number of major peaks are measured in the EI mode and matched by means of a computer with the intensity and mass of the major peaks of reference compounds compiled in spectral libraries. For structure elucidation, high-resolution mass spectrometers are used that are capable of measuring the mass of each ion to seven or eight significant figures. This enables the distinction between, for example, an OH group and an NH_3 group, both of which have 17 mass units but differ in the numerals beyond the decimal point. High-resolution mass spectrometers are usually attached to dedicated computers programmed to deduce the possible formula(s) from the mass measured. The highest mass, namely that of the molecular ion (M^+), gives the molecular formula of the compound, and the subsequent masses give the formulas of subsequent ions. The output of the computer is usually a printout showing the mass of each ion, starting with the highest, its intensity compared to the base (maximum intensity) peak, its possible structure(s), the molecular weight corresponding to each structure, and the deviation of the observed mass from the actual molecular weight. Figure 6 shows the computer printout for the high-resolution mass spectrum of a monosaccharide derivative.

Carbohydrate molecules present two difficulties to the mass spectrometrist: their low volatility, which prevents the recording of their spectra at low temperatures, and their instability toward the heat needed to volatilize them. Accordingly, in order to obtain a mass spectrum that shows the molecular ion and the subsequent molecular fragments, a free sugar must

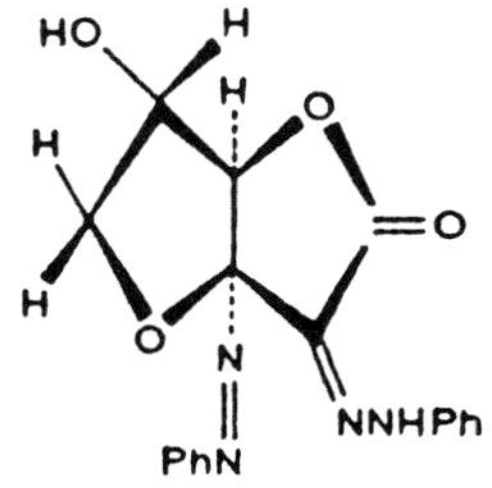

Measured Mass	Calculated Mass	Relative Intensity	Formula
354.1298	354.1328	0.36	$C_{18}H_{18}N_4O_4$
324.1130	324.1110	1.09	$C_{18}H_{16}N_2O_4$
249.0769	249.0786	1.70	$C_{10}{}^{13}C_2H_{11}N_2O_4$
248.0745	248.0746	14.05	$C_{11}{}^{13}CH_{11}N_2O_4$
247.0713	247.0719	100.00	$C_{12}H_{11}N_2O_4$
201.0667	201.0664	7.23	$C_{11}H_9N_2O_2$
188.0537	188.0501	1.05	$C_9{}^{13}CH_7N_2O_2$
187.0503	187.0508	8.35	$C_{10}H_7N_2O_2$
132.0450		1.06	
119.0605	119.0609	1.58	$C_7H_7N_2$
105.0442	105.0543	10.22	$C_6H_5N_2$
94.0623	94.0612	1.43	$C_5{}^{13}CH_7N$
93.0652	93.0578	16.18	C_6H_7N
92.0502	92.0500	61.90	C_6H_6N
91.0430	91.0422	5.63	C_6H_5N
78.0441	78.0425	6.23	$C_5{}^{13}CH_5$
77.0386	77.0391	61.30	C_6H_5
76.0307	76.0313	1.06	C_6H_4
66.0482	66.0469	1.04	C_5H_6
66.0428		1.71	
65.0401	65.0391	29.96	C_5H_5
64.0322	64.0313	3.95	C_5H_4
63.0239	63.0235	2.71	C_5H_3

Fig. 6. Portion of the printout obtained from a high-resolution mass spectrum of a monosaccharide derivative ($C_{18}H_{18}N_4O_4$) of molecular weight 354.1328.

either be converted into a volatile and stable derivative, such as an ether, an acetal, or an ester, that will withstand electron-impact mass spectrometry, or be studied in special instruments having chemical ionization or FAB capabilities, which are less accessible. In general, mass spectrometry has found less application in the study of monosaccharides and their derivatives than the previously discussed kinds of NMR spectroscopy. This is because the mass spectra of anomers, epimers, and stereomers are nearly all identical. On the other hand, mass spectrometry does offer insight into many useful structural features in carbohydrates. For example, because of the ease with which the C-1–C-2 bond in ketopyranoses is cleaved, they can be readily distinguished from aldopyranoses.

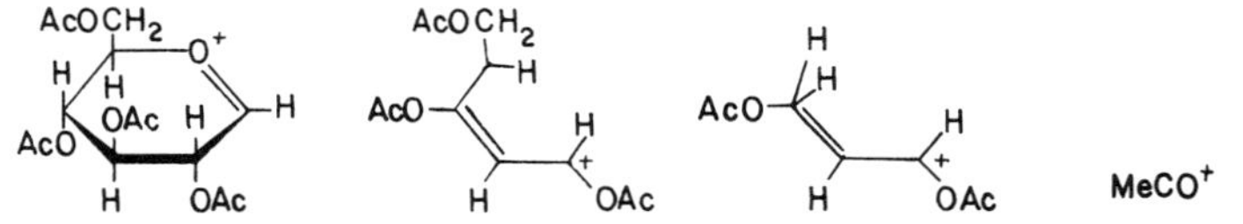
AcOCH2
O+
H
H
OAc H
H
AcO
H OAc
AcOCH2
H
AcO
H
+
H OAc
H
H
AcO
H
+
H OAc
MeCO+

m/z 331 *m/z* 242 *m/z* 157 *m/z* 43

β-D-Glucopyranose pentaacetate

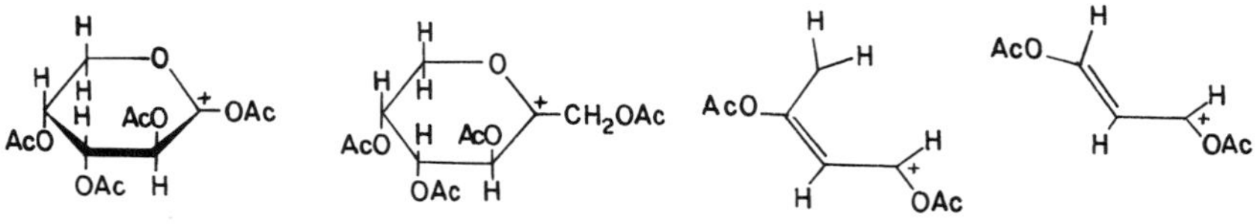
H
O
H
H
H AcO
+
OAc
AcO
OAc H
H
O
H
H
+
CH2OAc
AcO
H AcO
OAc H
H
H
AcO
H
+
H OAc
H
AcO
H
+
H OAc

m/z 317 *m/z* 331 *m/z* 170 *m/z* 157

β-D-Fructopyranose pentaacetate

For acyclic (aldehydo and keto) sugars, the cleavage occurs after the carbonyl group. Thus, peracetylated acyclic aldoses are cleaved between C-1 and C-2, and the corresponding ketoses are cleaved between C-2 and C-3.

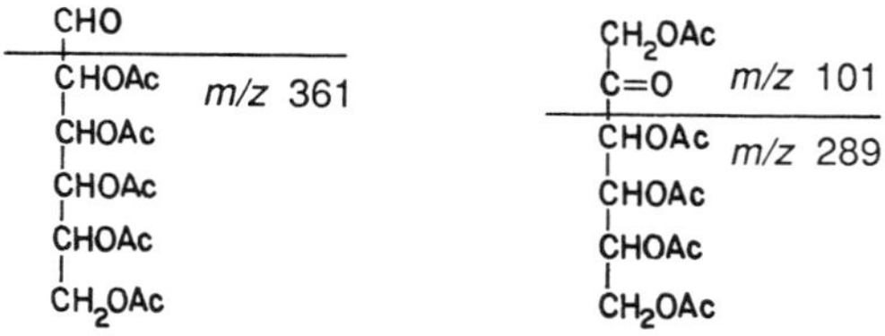
CHO
CHOAc
m/z 361
CHOAc
CHOAc
CHOAc
CH2OAc
CH2OAc
C=O
m/z 101
CHOAc
m/z 289
CHOAc
CHOAc
CH2OAc

B. Lindberg, in Sweden, has studied the mass spectra of methylated and acetylated alditols and used them in elucidating the structure of oligo-

and polysaccharides by subjecting the saccharides to methylation, reduction, and acetylation and then to gas chromatography–mass spectrometry. [For reviews see B. Lindberg and J. Lönngren, *Methods Enzymol.* **50,** 3 (1977); J. Lönngren and S. Svensson, *Adv. Carbohydr. Chem. Biochem.* **29,** 41 (1974).]

Acyclic sugar derivatives, such as dithioacetals, have characteristic fragmentation patterns that have also been used for structure elucidation. Thus, deoxy and aminodeoxy dithioacetals show moderately strong molecular peaks followed by three characteristic ions (A, B, C), which may be used to locate their deoxy or aminodeoxy functions. [See D. C. DeJongh and S. Hanessian, *J. Am. Chem. Soc.* **87,** 1408, 3744 (1965); **88,** 3114 (1966).]

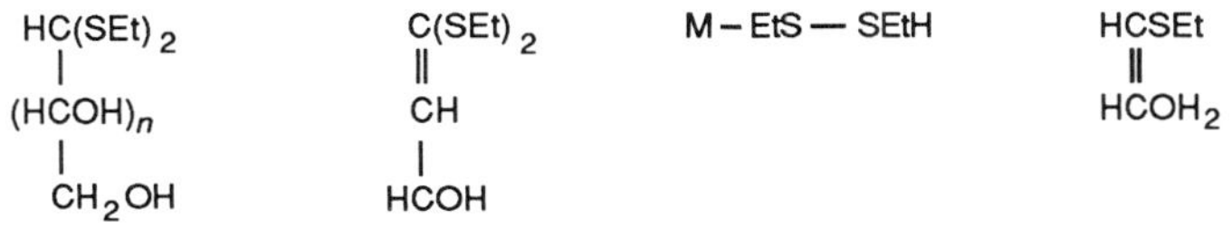

$HC(SEt)_2$–$(HCOH)_n$–CH_2OH $\quad$ $C(SEt)_2$=CH–HCOH (A) $\quad$ M – EtS — SEtH (B) $\quad$ HCSEt=$HCOH_2$ (C)

Pyranoses may be distinguished from furanoses because some of the carbonium ions resulting from the former differ from those of the latter.

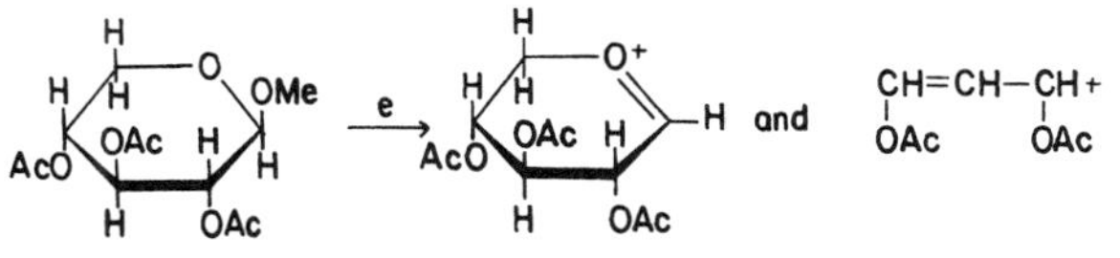

m/z 259 $\quad$ *m/z* 157

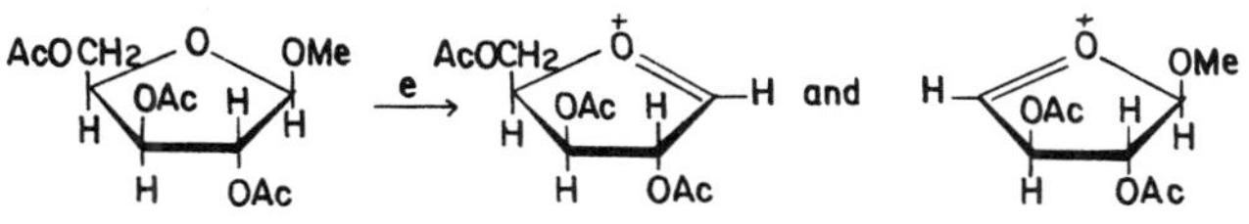

m/z 259 $\quad$ *m/z* 217

When the mass spectra of such monosaccharide derivatives as esters or isopropylidene acetals are inspected, it is important to recognize the carbonium ions associated with these groups. Thus, a monosaccharide acetate group will lose AcOH (*m/z* 60) and ketene (*m/z* 43), and isopropylidene acetals will lose a methyl group (*m/z* 15), to give stable five-membered carbonium ions.

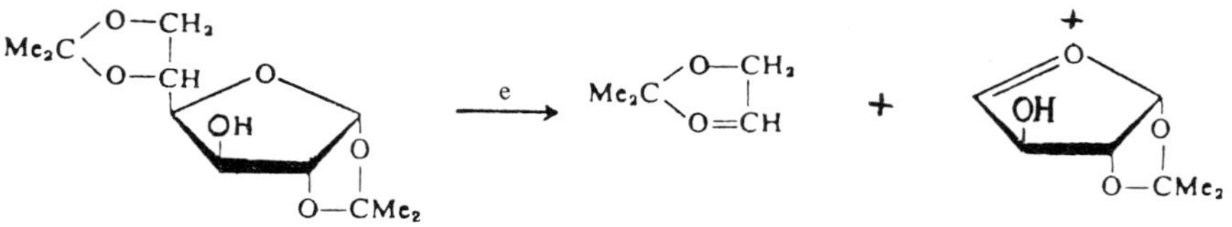

m/z 101 m/z 159

Nucleosides are fragmented between the base and the glycosyl group, as well as between C-1′ and C-2′ and between C-2′and C-3′.

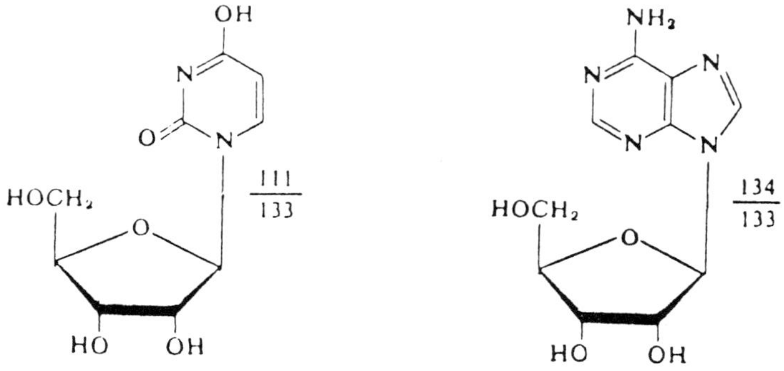

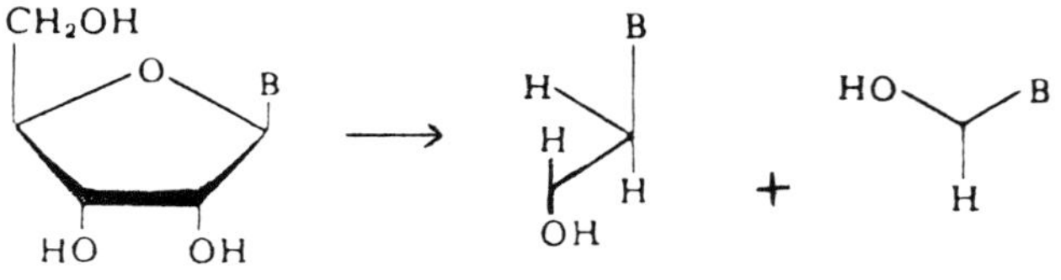

M − 89 B + 30

V. OPTICAL ROTATION, OPTICAL ROTATORY DISPERSION, AND CIRCULAR DICHROISM

The chirality of a molecule is established when its solution exhibits optical activity. This can be determined in three ways.

(1) By measuring its optical rotation with a polarimeter at one or more wavelengths. A beam of monochromatic polarized light is passed through a solution of the material, which, if chiral and optically pure, will cause the plane of the polarized light to bend either clockwise or counterclockwise. Because plane-polarized light has right- and left-circularly polarized components, which travel at different velocities in a chiral molecule, the

emerging light acquires a degree of ellipticity, and its plane of polarization is bent relative to that of the incident light. The sign (+ or −) and the extent of the bending, which is the angle measured in degrees between the planes of polarization of the incident and transmitted light, can be read directly in automatic instruments.

(2) By optical rotatory dispersion (ORD). The instruments in this case differ from those previously discussed in that the light source can emit vibrations extending over the whole UV and visible range. With every increment in the wavelength, the instrument measures the angle of rotation (ϕ) in degrees and records it on a graph versus wavelength (λ) in nanometers.

(3) By circular dichroism (CD). Instead of measuring optical rotations, as in the two previous techniques, the instruments here measure absorbance. Successive beams of right- and left-circularly polarized light are passed through the sample, which, if chiral, will retard one of them more than the other and produce elliptically polarized light. Because the instrument records the difference between two extinction coefficients (ε) at each wavelength, one due to the right-circularly polarized light and one to the left-circularly polarized light, the resulting CD curve is a plot of $\Delta\varepsilon$ versus wavelength (λ).

Both ORD and CD curves can exhibit marked changes in slope in the vicinity of the absorption maximum of a chromophore attached to the chiral center. This is known as the *Cotton effect* and is due to the rapid change in ellipticity near the chromophore absorption maximum, which is manifested by a rapid change in both ϕ and $\Delta\varepsilon$. Just before the absorption maximum is reached, ϕ undergoes an inversion in sign, which is visible in the ORD curve. In a CD curve, $\Delta\varepsilon$ continues its rise or fall until the absorption maximum is reached, and then its slope is inverted. The Cotton effect is said to be positive if ϕ or $\Delta\varepsilon$ increases with decreasing λ and negative if they decrease. Figure 7 shows the ORD and CD curves of a dextrorotatory compound that has a positive Cotton effect and of its levorotatory enantiomer that has a negative Cotton effect, together with their UV spectra.

A circular dichroism method has been developed that can determine the configuration of pyranose benzoates. A trisaccharide was permethylated, methanolyzed, and then benzoylated. One of the products showed a split Cotton effect at 253/238 nm. Since the methyl groups are transparent in that region, it was concluded that benzoate groups were responsible for the CD observed and that a glucopyranose with two vicinal equatorial benzoyl groups was present. [For more details see H.-W. Liu and K. Nakanishi, *J. Am. Chem. Soc.* **103,** 7005 (1981); N. C. Gonnella and K. Nakanishi, *J. Am. Chem. Soc.* **104,** 3775 (1982).]

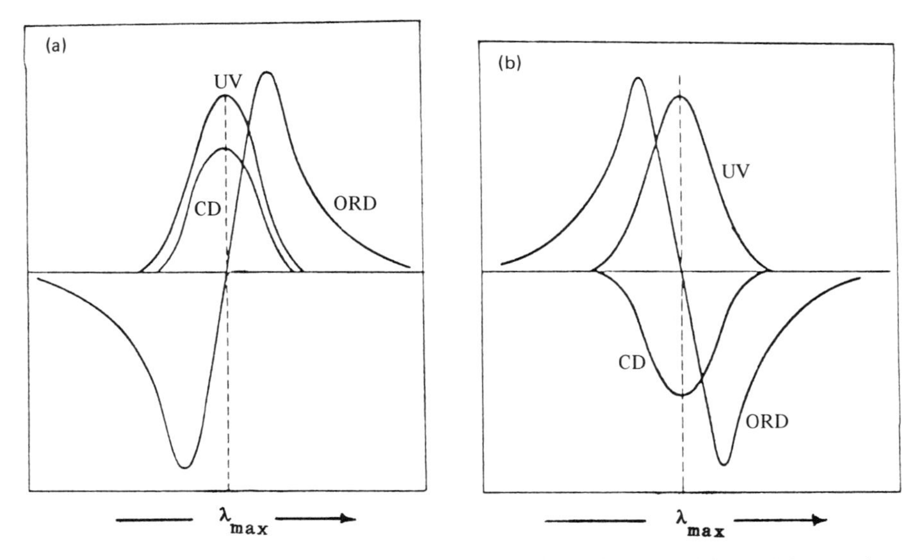

Fig. 7. Ultraviolet spectra of saccharide derivatives having (a) positive and (b) negative Cotton effects.

Most optical rotations recorded in the literature are reported either as *specific rotation,* $[\alpha]_D^{20}$, which is the rotation for a 1-g sample dissolved in 1 mL of solution, placed in a cell 1 dm (10 cm) long, measured at room temperature (20°) at the wavelength of the sodium D line (589 nm), or as *molecular rotation,* which is $[\alpha]_D^{20} \times$ molecular weight/100. Because of the availability of a large number of specific and molecular rotations measured at the wavelength of the sodium D line, which was used as a standard by Hudson, carbohydrate chemists have continued to record such measurements despite the advent of ORD and CD instruments. The latter gives valuable information in the case of complex molecules that show multiple Cotton effects.

Specific rotations are useful for characterizing new derivatives and recognizing known ones. They are also of value in sugar analysis, especially for purity control, or as a means of determining the concentration of solutions; for example, saccharimeters are widely used in the sugar industry to determine the concentration of syrups. In structural work, which is the concern of this chapter, polarimetric, CD, and ORD measurements have been used to determine the configuration of unknown saccharide derivatives, as will be seen next.

A. Rules Correlating Optical Rotation with Configuration

Ever since van't Hoff and Le Bel discovered that asymmetric carbon atoms are responsible for the optical activity of (certain) organic compounds in solution, chemists have attempted to correlate the sign and magnitude of rotation with the configuration of the chiral centers. Carbohydrates constitute a group of compounds that contain several similarly substituted asymmetric carbon atoms, which are usually attached to H atoms, OH groups, or other carbon atoms that are, in turn, attached H atoms and OH groups. Free sugars usually exist in cyclic forms that do not possess chromophores, and their sign of rotation is greatly influenced by the configuration of the carbon atoms attached to the ring oxygen atom. On the other hand, some saccharide derivatives are acyclic and therefore do not have a ring oxygen atom. In such compounds, the signs and magnitude of rotation are influenced mainly by the configuration of the chiral center attached to an unsaturated chromophore.

1. Effect of Configuration of Centers Attached to the Ring Oxygen Atom

Most free sugars do not possess unsaturated chromophores, because they exist preponderantly in cyclic forms. Their highest absorptions are due to the $n \rightarrow \sigma^*$ transitions of the ring oxygen atom, which occur at 200 nm. In such cyclic compounds, the chiral carbon atoms attached to the ring oxygen atom affect the overall rotation significantly more than the remaining ring carbon atoms. Thus, for example, C-1 and C-5 have the greatest influence on the rotation of a pyranoside. This readily explains the rule, developed by Hudson, that *in cyclic sugars and glycosides, the more dextrorotatory of an anomeric pair in the* D *series* (*the* α *form*) *has its C-1 OH to the right in a Fischer projection, and vice versa.* This rule holds not only for pentofuranosides and hexopyranosides, where both C-1 and the chiral center that determines the D and L notation are attached to the ring oxygen atom, but also for higher sugars, because a change from a D to an L configuration in a given sugar will bring about an inversion in the configuration of all of the chiral centers, including the carbon atoms attached to the ring oxygen atom. This is why the α-D anomer of a sugar (the more dextrorotatory of the α, β pair in the D series) is the enantiomer of the corresponding α-L form (the more levorotatory of the α, β pair in the L series).

The effect of the configuration of the chiral centers attached to the ring oxygen atom is apparent both in the absence of unsaturated chromophores, as already shown, and in the presence of such chromophores, as will be seen next.

2. Derivatives Having Chiral Centers Attached to Unsaturated Chromophores

The sign of rotation of a sugar derivative possessing unsaturated chromophores capable of undergoing $\pi \rightarrow \pi^*$ or $n \rightarrow \pi^*$ transitions, such as C=C, C=O, or C=N groups or aromatic or heteroaromatic rings, is greatly influenced by the configuration of the chiral center attached to such chromophores. Thus, in the case of acyclic sugar derivatives (which do not have a ring oxygen), the sign of rotation is determined solely by the configuration of the center attached to the unsaturated chromophore, irrespective of the configuration of the other centers. This rule is useful because it permits the determination of the configuration of a chiral center attached to a chromophore, not only in the sugar derivative but also in the corresponding the parent sugar. Thus, in aldonic acid amides, hydrazides, and quinoxalines, the center attached to the unsaturated chromophore corresponds to C-2 of the parent sugar, and in osotriazoles, to C-3 of the parent sugar. Accordingly, the configuration of C-2 and C-3 in a sugar can be established by measuring the rotation of these derivatives. Rotation rules have been put forward by Hudson for hydrazides, Richtmyer for quinoxalines, and El Khadem for osotriazoles. These were later combined by the author in a generalized rule, applicable to acyclic sugar derivatives having unsaturated chromophores. The generalized rotation rule states that *the sign of rotation of an acyclic saccharide derivative (a polyhydroxyalkyl chain) having one of its chiral carbon atoms attached to an aromatic or heteroaromatic ring is determined solely by the configuration of the chiral center attached to the chromophore; it is dextrorotatory if this chiral center has the* D *configuration and levorotatory if it has the* L *configuration in a Fischer projection.*

This rule applies to acyclic saccharide derivatives because they do not possess a furanose or pyranose ring (which enhances the effect of the groups attached to their ring oxygen atom). The rule also applies to several hydroxyalkyl chains attached to unsaturated acyclic chromophores. These include the amides and hydrazides of aldonic acids, as well as alkali metal carboxylates and nitriles. It does not apply to rings attached to *substituted* hydroxyalkyl chains, because esters, ethers, cyclic acetals, etc. enhance the hierarchy of the center(s) to which they are attached and lead to exceptions.

In the case of cyclic sugar derivatives having unsaturated chromophores, such as lactones, the chiral center attached to the ring oxygen atom will add its effect to that of the center attached to the chromophore. This is illustrated in Hudson's lactone rule, which states that *if the OH group involved in lactonization, i.e., whose oxygen atom will become part of the ring (O-4 in a γ-lactone and O-5 in a δ-lactone), is to the right in a*

Fischer projection, the rotation of the resulting lactone will be more positive than that of the free acid, and vice versa.

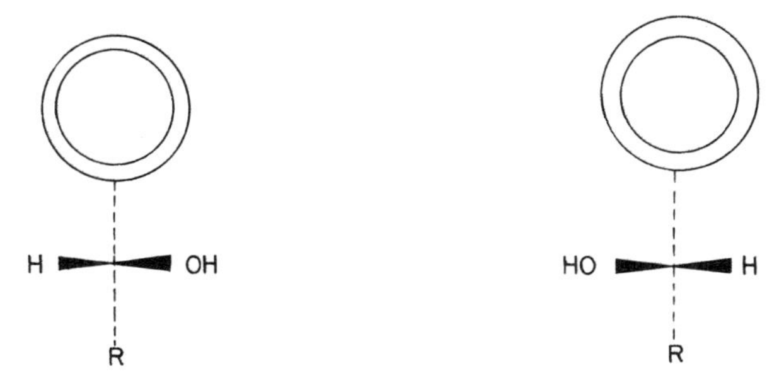

Aromatic or heteroaromatic ring attached to hydroxyalkyl chain having positive rotation

Configuration of same system when rotation is negative

B. Empirical Methods of Calculating the Sign and Magnitude of Rotation

If it is assumed that the contribution of each chiral center to the overall rotation is not affected by the configuration of the other chiral centers, it is possible by simple arithmetic to determine the contribution of the anomeric chiral center (or, for that matter, any other center) to the overall rotation and the contribution of the remaining chiral centers to the overall rotation. Thus, by subtracting the specific rotation of the α and β anomers of different sugars and dividing the result by 2, Hudson was able to show, in his *isorotation rule*, that the contribution of the anomeric center in glycopyranoses amounts to about +42°. Later, Isbell determined the contribution of the remaining chiral centers by adding the rotations of α and β anomers and dividing by 2, and obtained similar results for identical configurations.

C. Effect of Conformation

Once the favored conformation of a sugar has been determined by one of the methods described in Chapter 2, it is possible to use empirical methods to predict its molecular rotation. Whiffen, Brewster, and Lemieux each developed a method for calculating the rotation of cyclic sugars that takes into consideration the spatial arrangement of the substituents attached to a ring in a given conformation. For example, Lemieux summed numerical values given to pairs of oxygen atoms or of carbon and oxygen atoms attached to contiguous atoms. Then he assigned signs to these values depending on whether, in subsequent Böeseken ("Newman") projections that start with C-1 (the latter positioned so that it is nearest the viewer), the oxygen atom farthest from the viewer is to the right (+) or to the left (−) of the oxygen atom nearest the viewer. Lemieux assigned to two oxygen atoms attached to adjacent carbon atoms a rotation value of 55°. However, if one of these oxygen atoms forms the ring, the value is increased to 90°; a carbon–oxygen interaction is given a value of 45°, and an oxygen–carbon interaction across the ring oxygen atom a value of 115°.

Although all of the foregoing computations give satisfactory results with cyclic sugars and their methyl glycosides, they do not apply to derivatives in which the anomeric carbon atom is linked directly to atoms other than oxygen. In these cases, the sign of rotation and the sign of the Cotton effect vary from one derivative to another. Thus, for example, the signs of rotation of β-D-ribofuranosyl groups attached to different achiral groups are by no means the same. They vary according to the favored orientation of the achiral groups around the pivot bond, which changes the direction of the dipole moment of the aglycon vis-à-vis that of the C-1 → O-4 vector in the sugar. Because the conformation of the most stable rotamer depends on the nature of the achiral group attached to the sugar, the sign of rotation will change from one group to another. This is best illustrated as in Fig. 8, which depicts the signs of rotation of some nucleosides having β-D-ribofuranosyl groups attached to purines and pyrimidines at different positions of the base. For example, β-D-furanosyl rings attached to N-9 of purines have negative rotations and those attached to position 7 have positive ones (see Fig. 9).

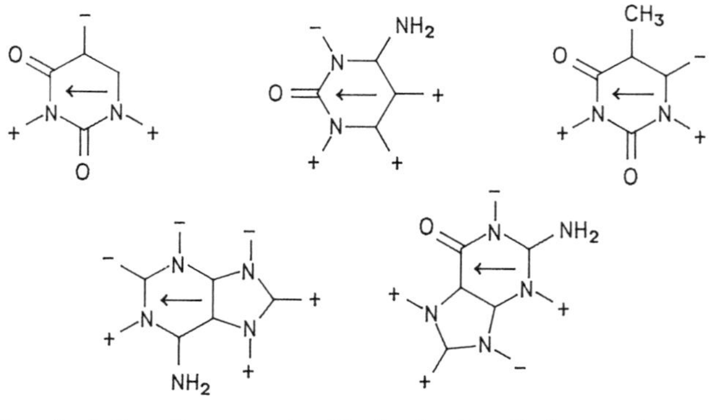

Fig. 8. Sign of rotation of some nucleosides having β-D-ribofuranosyl groups attached to bases. The arrows point to the direction of the dipole moment of the base. For examples see Fig. 9.

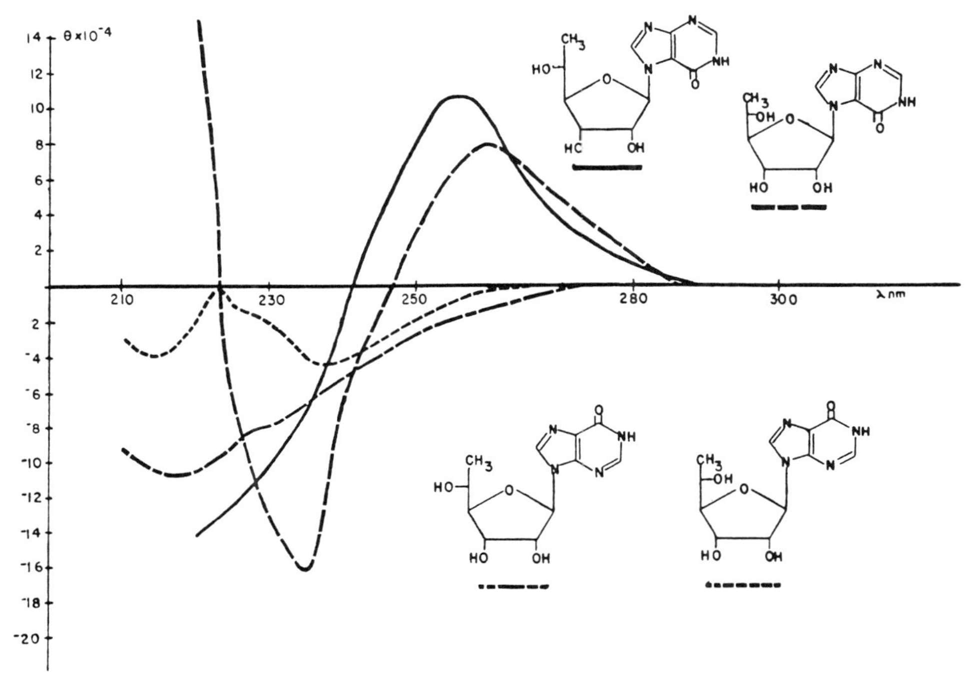

Fig. 9. CD spectra of 6′-deoxy-β-D-hexofuranosylhypoxanthines substituted at N-9 (− Cotton effect) and at N-7 (+ Cotton effect).

D. Mutarotation

Mutarotation is defined as the change in the observed optical rotation with time. It is the outcome of the formation in solution of one or more isomers of the starting material whose contribution to the end rotation depends on the rotation of each isomer and its concentration. If the change in rotation plotted against time follows first-order reaction kinetics, it may be concluded that only two main components are in equilibrium, and the mutarotation is designated a simple mutarotation. Any deviation from first-order kinetics indicates that at least three components are present in equilibrium, and the mutarotation is designated complex. Because, as discussed in Chapter 2, two pyranoses, two furanoses, and two acyclic forms can exist in equilibrium in solution, the fact that α-D-glucopyranose exhibits a simple mutarotation suggests that only the α- and β-pyranoses exist in solution in appreciable proportions.

Because the opening of a cyclic form of a sugar involves protonation of the ring oxygen atom and extraction of the anomeric hydroxyl proton by base, mutarotation is promoted both by acids and bases. The rate of reaction is also increased by increasing temperature, which may double or triple the reaction rate with every increment of 10°C.

It should be noted that in strongly basic media free sugars undergo enolizations, isomerizations, and degradations. Because these reactions are often irreversible, the change in rotation that accompanies them is not true mutarotation.

PROBLEMS

1. Depicted below are three ^{1}H-NMR spectra and three formulas. Match the spectra with the structures.

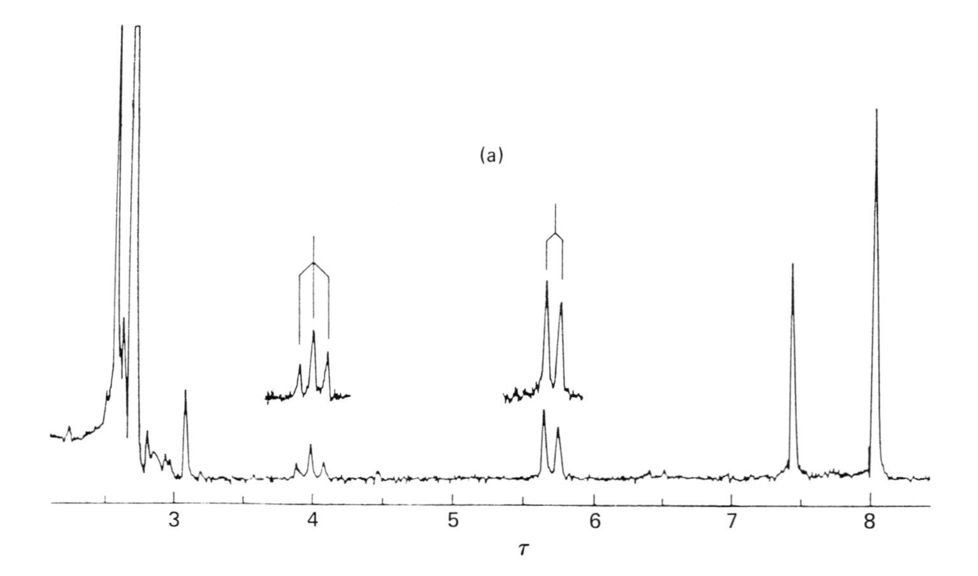

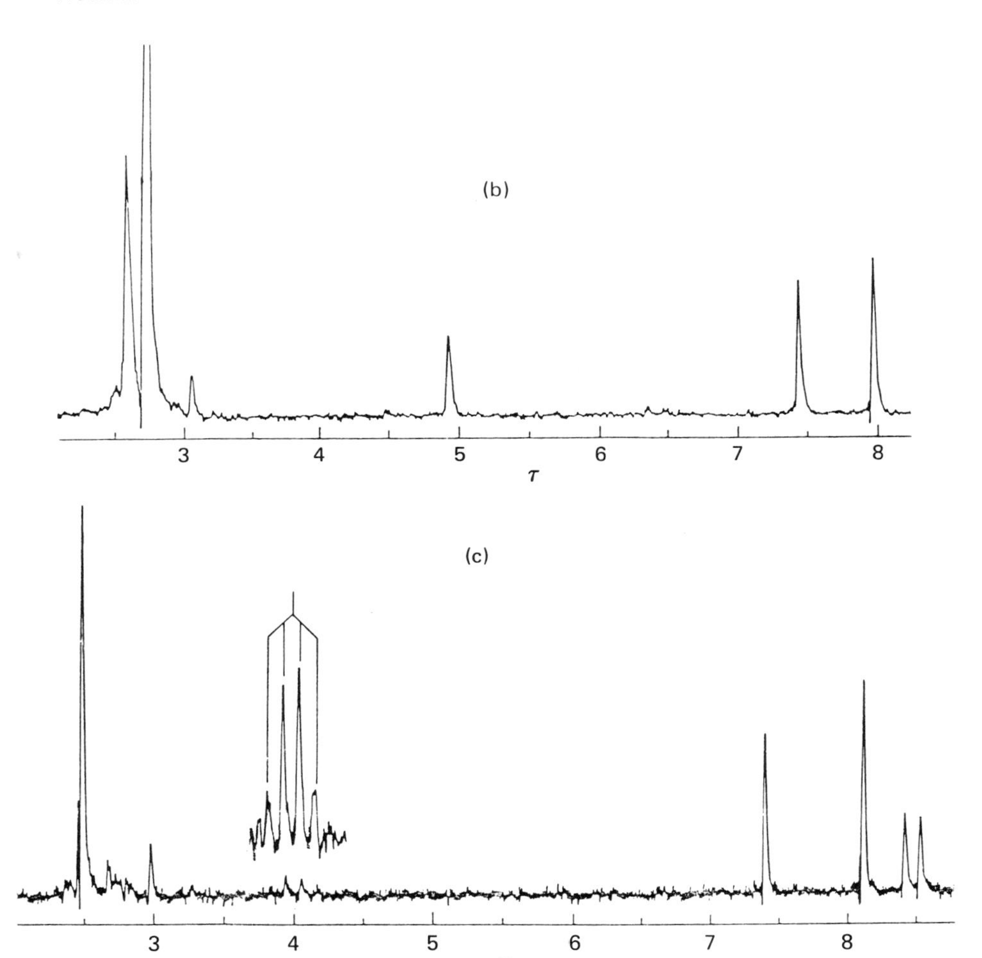

1

2

3

2. The following ^{1}H-NMR spectra were measured for three 2,6-dideoxyhexopyranosyl derivatives: (a) di-O-acetyl-2,6-dideoxy-α-L-*lyxo*-hexopyranosyl bromide; (b) di-O-acetyl-2,6-dideoxy-α-L-*arabino*-hexopyranosyl bromide; (c) 2,6-dideoxy-β-D-*ribo*-hexopyranose. Use the H-4 coupling to determine the ring conformation.

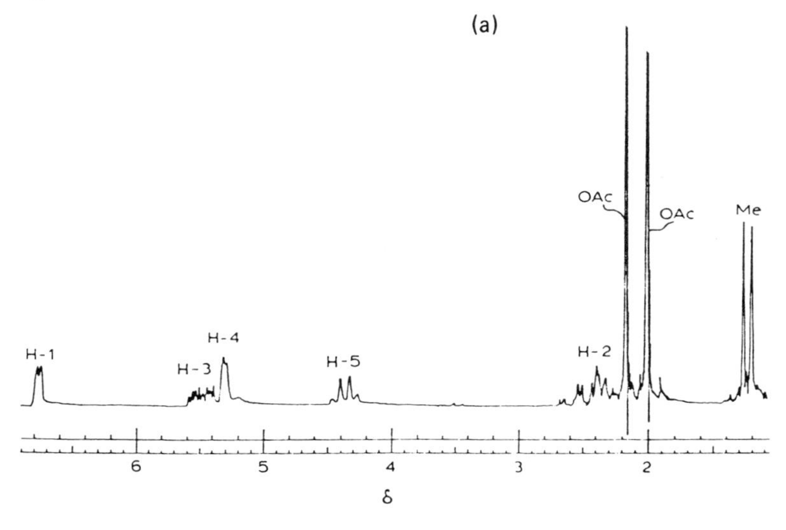

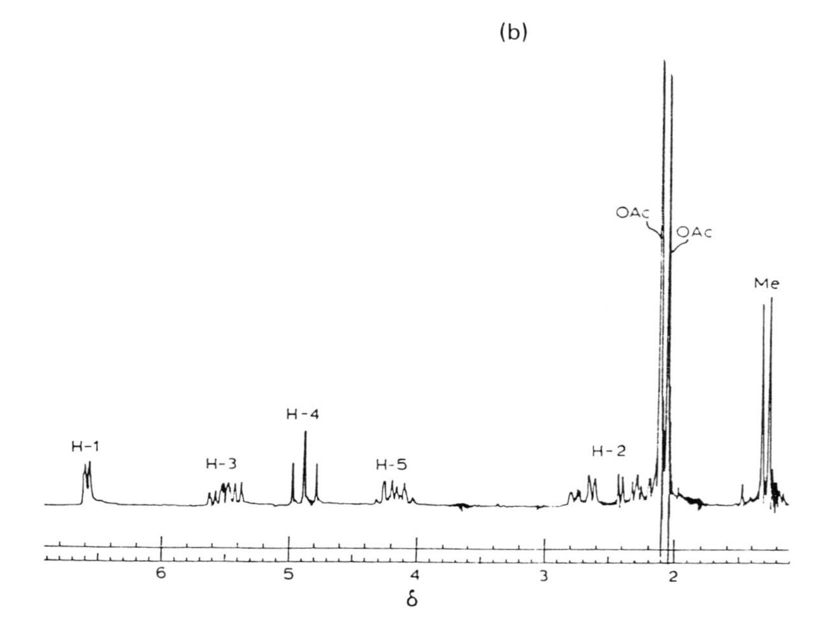

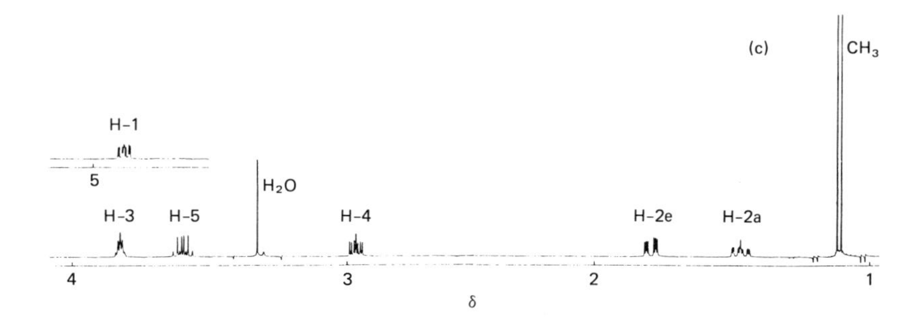

3. The ^{1}H-NMR spectra of (a) peracetylated D-xylose; (b) D-glucose; and (c) D-galactose are shown. Deduce the anomeric configuration and ring conformation.

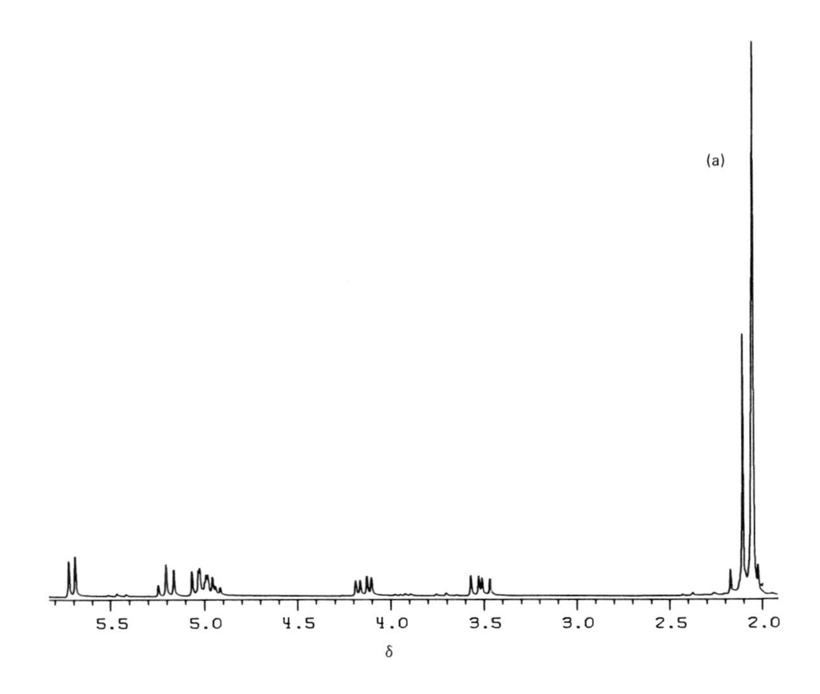

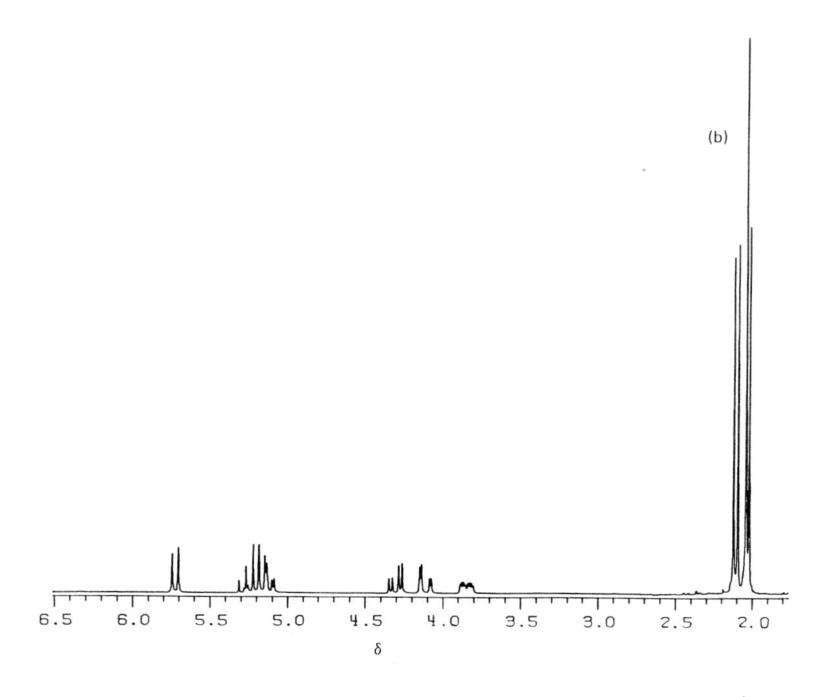
(b)
6.5
6.0
5.5
5.0
4.5
4.0
3.5
3.0
2.5
2.0
δ

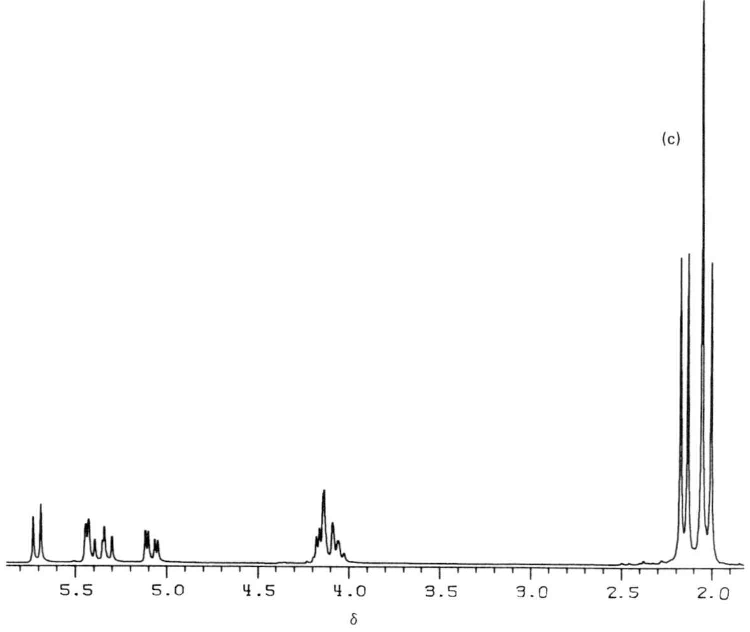
(c)
5.5
5.0
4.5
4.0
3.5
3.0
2.5
2.0
δ

4. The NMR spectra shown depict peracetylated hexitols. One is a cyclic meso compound, *myo*-inositol, and two are acyclic alditols, glucitol and mannitol. Match the spectra.

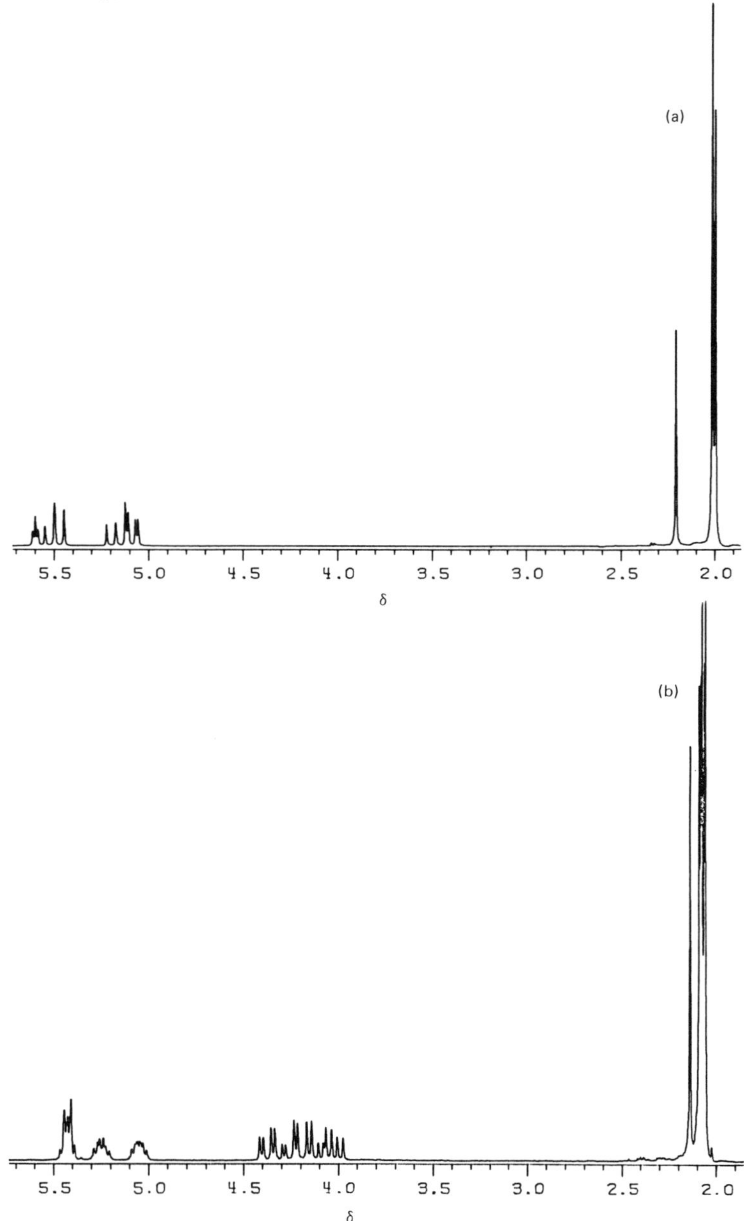

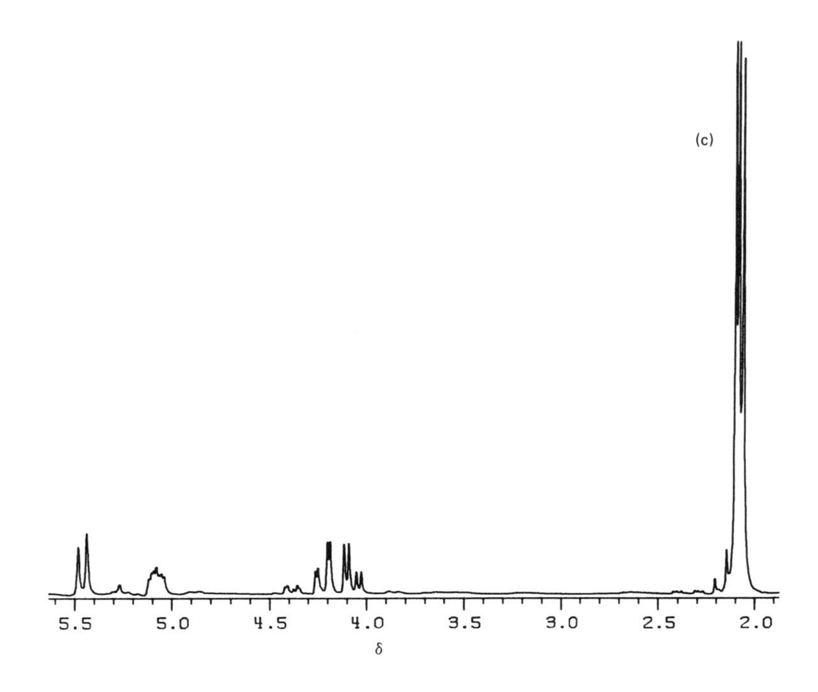

5. The following major peaks were observed in the mass spectrum of the nucleoside antibiotic cordycepin: (a) *m/z* 135 (100%); (b) *m/z* 164 (50%); (c) *m/z* 178 (12%); (d) *m/z* 221 (12%); (e) *m/z* 251 (10%). Assign structures to peaks (a)–(e).

6. The mutarotation (measured rotation minus rotation at equilibrium) of β-D-glucose in water is depicted below. Draw the structures responsible for the observed kinetics.

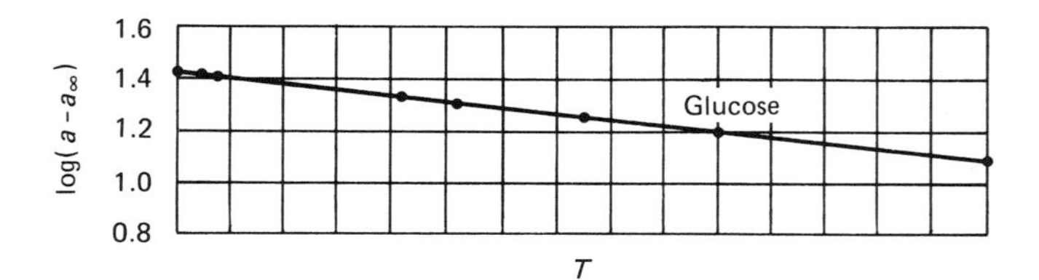

5

Reactions of Monosaccharides

I. ISOMERIZATION

Many monosaccharides exist in solution as equilibrium mixtures composed of two acyclic forms (the aldehydo form and its hydrate) and at least four cyclic structures (two furanoses and two pyranoses). Conversion of any one of these isomeric forms into the others can be achieved in a neutral medium by what is referred to as *mutarotation* (or anomerization) *reactions*. Further isomerizations occur in alkaline medium and lead to (a) the conversion of aldoses to the epimeric aldoses and the related ketoses or of ketoses to the epimeric aldoses by the *epimerization reaction*; (b) the formation of higher epimeric aldoses and ketoses; and finally (c) chain rupture by reverse-aldol reactions and subsequent recombination to yield rearrangement products. It should be noted that all these reactions occur concurrently, but their rates depend on the type of saccharide, the pH (concentration of the alkali), the nature of the cation, and the duration of the reaction.

A. Mutarotation Reactions (Anomerization Reactions)

The currently accepted mechanism of mutarotation reactions assumes that the interconversion of the pyranose and furanose forms proceeds through acyclic intermediates which retain the shape (conformation) of the rings from which they were formed. These acyclic intermediates are called pseudoacyclic structures and exist in four forms (two from the α and β pyranoses and two from the α and β furanoses).

In neutral or acidic solutions, sugars undergo rapid protonation on the ring oxygen atom, followed by slower movement of electrons from the anomeric hydroxyl group to release the ring oxygen. This results in the formation of one of the pseudoacyclic forms. Rotation of C-1 of the pseudoacyclic intermediate 120° gives the epimeric pseudoacyclic intermediate, which by ring closure gives the anomer of the same form, whereas rotation of C-4 120° would give a pseudoacyclic precursor of the other form. In alkaline media the reaction is initiated by transfer of the anomeric proton to the base, followed by hydration to form pseudoacyclic intermediates, which by the reverse process would yield the anomeric pyranose.

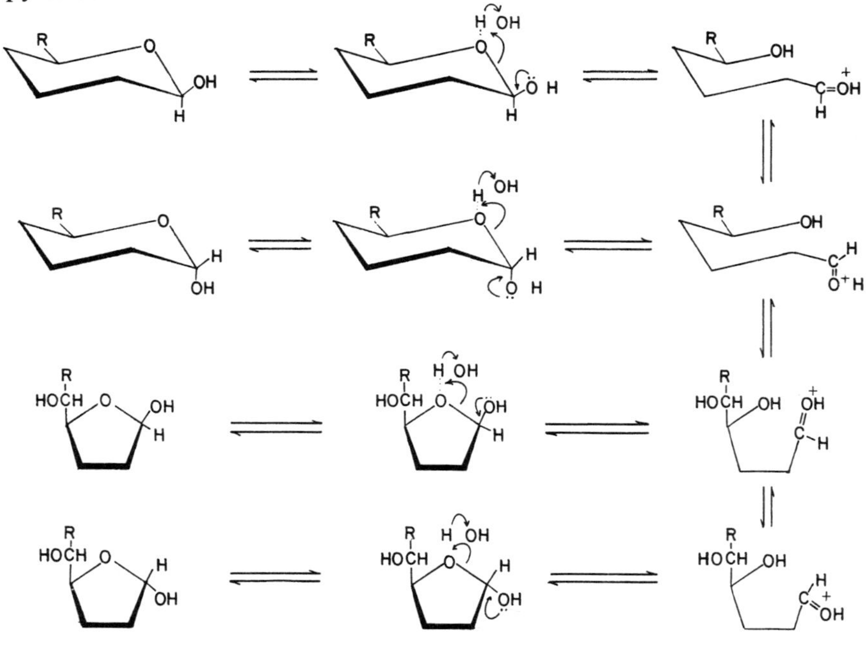

B. Epimerization Reactions

Epimerization reactions occur in alkaline media and result in the interconversion of two aldoses and a related ketose to give an equilibrium mixture of the three. Two mechanisms are responsible for these results; the predominant one, *reversible enolization,* is sometimes accompanied by isomerization by *hydride shifts.* Both mechanisms start with the acyclic form of the sugar produced by the mutarotation reaction. The first mechanism is initiated by abstraction of a proton α to the carbonyl group, while the second starts with ionization of the proton of the α hydroxyl group. Migration of electrons leads to an ionized enediol in the first case and to an ionized cyclic hydride intermediate in the second. Further electron shifts will lead to the epimeric aldose and related ketose. It should be noted that in the second mechanism there is no exchange of protons with the medium, whereas protons are exchanged in epimerizations by reversible enolization.

Epimerization by reversible enolization:

```
  HC=O     -H+     HC=O             H-C-O-             HC=O    +H+     HC=O
   |       ⇌        |        ⇌        ||        ⇌        |      ⇌        |
H-C-OH     +H+    -C-OH              C-OH           HO-C-      -H+   HO-C-H
   |                |                  |                 |               |
                                       ⇅
   H
  HC-OH    +H+    H-C-O            H-C-OH
   |       ⇌        |        ⇌        ||
  C=O      -H+     C=O               C-O-
   |                |                  |
```

Epimerizations by hydride shifts:

```
H-C=O               H  ┐-            H
  |                 C-O             H-C-O-
  |        ⇌     H′ |        ⇌        |
H-C-O-              C-O              C=O
  |                 |                 |
```

[See G. B. Gleason and R. Barker, *Can. J. Chem.* **49**, 1433 (1971).]

C. Fragmentation Reactions

These reactions occur in alkaline media and result in breaking of the saccharide chain by the reverse-aldol reaction. The reaction is initiated by ionization of the hydroxyl group in the β position with respect to the carbonyl group. This is followed by formation of a carbonyl group and breaking of the bond between the α and β carbons, yielding an enediol, which will isomerize to the more stable keto form. The two fragments thus formed can recombine by the aldol reaction to form the 3- and 4-epimers of the starting sugar.

It should be noted that the formation of higher epimers (3-epimers, 4-epimers, etc.) can also take place without fragmentation, by the reversible enolization pathway or by hydride shifts. Table I shows the characteristics of isomerizations by the different pathways.

TABLE I

Characteristics of Isomerizations by Reversible Enolization, by Hydride Shifts, and by Fragmentation–Recombination

	Reversible enolization	Hydride shifts	Fragmentation and recombination
Order of formation of 2-, 3-, and 4-epimers:			
	2- then 3- then 4-epimers	2- then 3- then 4-epimers	Simultaneous
Isomers belonging to other series from:			
Aldohexoses:	Not formed	Not formed	Formed
Hexuloses:	Not formed	Not formed	Not formed
Aldopentoses:	Not formed	Not formed	Formed
Number of hydrogen atoms exchanged in alkali:			
Aldohexoses:	1	None	3
Hexuloses:	3	None	5
Aldopentoses:	1	None	3
Pentuloses:	3	None	3
Distribution of carbon label in isomers obtained from C-1 labeled sugars:			
Aldohexoses:	C-1 and C-6	C-1 and C-6	C-1, C-2
Hexuloses:	C-1 and C-6	C-1 and C-6	C-1, C-3, C-4, C-6
Aldopentoses:	C-1 and C-5	C-1 and C-5	C-1, C-2
Pentuloses:	C-1 and C-5	C-1 and C-5	C-1, C-3

II. ADDITION AND SUBSTITUTION REACTIONS

In this section the reactions characteristic of the carbonyl group will be discussed first, starting with reactions involving condensation or addition of nucleophiles (that is, with or without loss of water). This will be followed by a study of the reaction products obtained by oxidation and by reduction of this highly reactive group. Finally, the enolization of the carbonyl group will be discussed. The reactions of the hydroxyl groups will then be treated, starting with nucleophilic substitution reactions (S_N1

or S_N2) in which a nucleophile attacks a carbon atom bearing a hydroxyl group and displaces the latter. This will be followed by reactions in which the oxygen atom of the hydroxyl group acts as a nucleophile, adding to carbonyl groups or displacing leaving groups. The chapter will conclude with a discussion of the products formed by the oxidation of various hydroxyl groups.

A. Reactions of the Carbonyl Group

Free sugars capable of cyclization exist in solution as equilibrium mixtures containing very little of the acyclic (aldehydo or keto) form, and yet they react with carbanions to give high yields of the addition products of the acyclic form. As soon as the initial amount of acyclic form is consumed, more is produced by decyclization, and so on, until all of the cyclic form has been converted into the acyclic form and has reacted with the carbonyl reagent.

Although addition of a carbon, nitrogen, oxygen, or sulfur nucleophile to the carbonyl group of a monosaccharide gives acyclic products, these may later cyclize. On the other hand, intramolecular nucleophilic addition, as by an OH group attached to the sugar chain, on the carbonyl carbon atom can only give cyclic products.

The reactions discussed in this section will start with the nucleophilic addition of carbon, nitrogen, oxygen, and sulfur to the carbonyl carbon atom. This will be followed by a review of the oxidation and reduction products of the acyclic forms of sugars. Finally, reactions involving enolization of the carbonyl group and migration of the double bonds will be discussed.

1. Reaction with Carbon Nucleophiles

Addition of carbon nucleophiles to the carbonyl group of aldoses has been widely used to extend the carbon chains of saccharides, i.e., to ascend the series. Similar additions to the keto group of glyculoses have been used to prepare branched sugars. Of particular value in forming C–C bonds are the reactions of the nucleophiles ^{-}CN, $^{-}CH_2NO_2$, and $^{-}CH_2N_2$ with free sugars as well as those of ylides and organometallic nucleophiles used in Wittig and Grignard reactions with protected sugars.

a. *Addition of HCN, the Cyanohydrin Synthesis.* The cyanohydrin synthesis is the oldest method used in carbohydrate chemistry for ascending the series. It was first developed by Kiliani and later extensively applied by Fischer in his structure elucidation of aldoses (see Chapter 2). The reaction is initiated by the addition of aqueous NaCN (this is preferred to

HCN, which is more difficult to handle) to a cold aqueous monosaccharide solution. Because a new chiral center is introduced into the molecule, two epimeric nitriles are formed, whose ratio was found by Isbell to depend on the pH of the medium. The nitriles are hydrolyzed when the reaction mixture is heated. The resulting ammonium salts are separated or converted into difficultly soluble salts, converted into the aldonic acid lactones (not shown in the scheme), and then reduced with sodium amalgam to give the higher aldoses.

$^{-}$CN + HC=O →

H—C(CN)—OH → H—C(COONH$_4$)—OH → → H—C(HC=O)—OH

HO—C(CN)—H → HO—C(COONH$_4$)—H → → HO—C(HC=O)—H

The reverse of the cyanohydrin synthesis occurs in the Wohl degradation, where the oxime of a monosaccharide is refluxed with acetic anhydride to give the acetylated cyanohydrin. On treatment with ammonia, this is first hydrolyzed to the acetylated lower saccharide, which then forms an addition product with ammonia. The latter was found by Isbell to undergo migration of the acetyl groups from O to N and to give a 1,1-bis(acetamido)-1-deoxyalditol, which is isolated at this stage. When this compound is subjected to acid-catalyzed hydrolysis, it readily yields the lower aldose.

HC=NOH / H—C—OH / H—C—OAc →(NH_2OH) CN / H—C—OAc / H—C—OAc →(Ac_2O) HC=O (+ NH$_3$) / H—C—OAc →(NH_3) NH$_2$ / H—C—OH / H—C—OAc → HC(NHAc)$_2$ / H—C—OH →(H^+) HC=O / H—C—OH

b. *Addition of Nitroalkanes (Nef Reaction).* Addition of nitroalkanes to monosaccharides is usually conducted in basic media (NaOH or NaOMe) in order to abstract the α proton of the nitroalkane and form the reactive nucleophile. The reaction is exemplified by the addition of the carbanion generated from nitromethane to the carbonyl group of an acyclic aldose, to form two epimeric deoxynitroalditols having one carbon atom more than the starting sugar.

To complete the Nef reaction, the nitroalditols are converted into salts with bases, and these are hydrolyzed with acid to give two epimeric higher aldoses. The reaction of nitroalkanes with sugars was extensively studied by H. H. Baer.

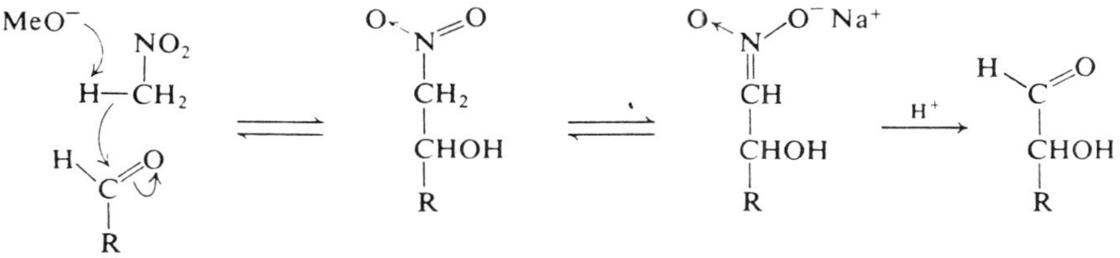

c. *Addition of Diazomethane.* Diazomethane adds to the carbonyl group of protected aldehydo sugars to form the 1,2-epoxide of the higher alditol of the series.

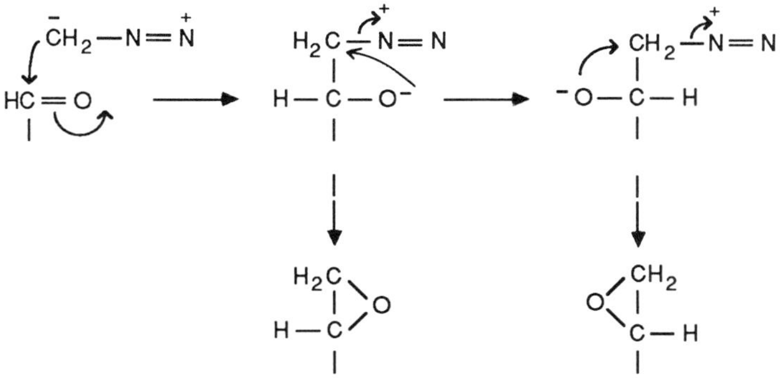

Wolfrom and co-workers used diazomethane to prepare higher ketoses from aldonic acid chlorides by nucleophilic acyl substitution. They found that blocked aldonic acid chlorides react with diazomethane to give diazo derivatives, which, on hydrolysis, give ketoses having one carbon atom more than the acid chloride. It may be noted that there is a difference between this reaction and the Arndt–Eistert reaction, where rearrangement of the substitution product occurs via a carbene intermediate.

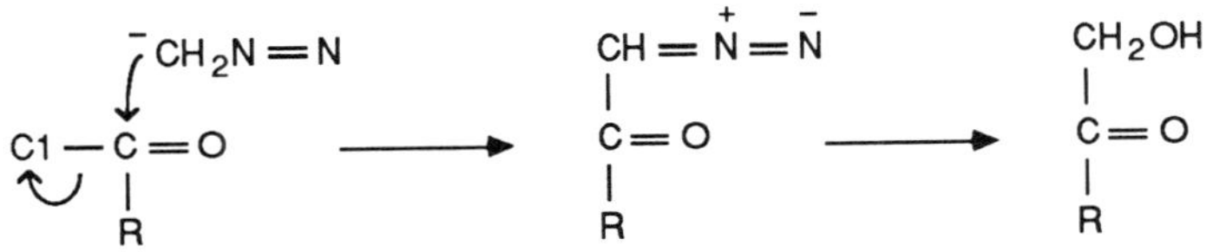

d. *Addition of Ylids, the Wittig Reaction.* The Wittig reaction may be performed on O-protected or nonblocked acyclic or cyclic sugars. To achieve this, a phosphonium salt, prepared beforehand from an alkyl halide and phosphine, is caused to react with a base (usually BuLi) to generate the ylid, which then adds to the carbonyl group to give an alkene. The reaction can be performed with stabilized ylids as well.

$$Ph_3P + ICH_3 \longrightarrow Ph_3\overset{+}{P}\text{-}CH_3 + BuLi \longrightarrow Ph_3\overset{+}{P}\text{-}\overset{-}{C}H_2 + HC{=}O \longrightarrow HC{=}CH_2$$

e. *Addition of Organometallic Carbanions.* Like the Wittig reaction, the addition to carbohydrates of a Grignard reagent and such other organometallics as metal acetylides necessitates the presence of a sufficient amount of the free carbonyl form. Such reactions are not performed on free sugars. Furthermore, because hydroxyl groups possess active hydrogen atoms that may react with the organometallic compounds, they must be protected with groups other than esters. An example of a chain extension achieved by means of a Grignard reaction is the conversion of the aldehyde obtained by oxidation of methyl 2,3-O-isopropylidene-β-D-ribofuranoside into two 5-epimeric 6-deoxyhexose derivatives. The reaction depicted below involves a C-5 carbonyl group and illustrates an alternative route to ascend the series [see H. El Khadem and V. Nelson, *Carbohydr. Res.* **98,** 195 (1981)]. If one epimer is desired, the stereoselectivity may be enhanced by complexation with Lewis acids such as $TiCl_4$ or BF_3OEt_2 [for examples see M. T. Reetz *et al., Angew. Chem.; Int. Ed. Engl.* **22,** 725 (1983); **23,** 556 (1984)].

2. Reaction with Nitrogen Nucleophiles

The nucleophilicity of nitrogen is enhanced if an adjacent atom has one or more unshaired pairs of electrons. This is called the *alpha effect* and is the reason why NH_2NH_2 and NH_2OH are stronger nucleophiles than NH_3.

The nitrogen nucleophiles commonly used with monosaccharides contain a primary amino group attached to a carbon, a nitrogen, or an oxygen atom. Products obtained from the first type of nucleophile are rarely of the acyclic Schiff base type, because they readily cyclize. The other two types of nucleophiles give carbonyl group derivatives that usually exist in acyclic forms. The major pathway for the reaction between a nitrogen nucleophile and a free sugar is via the acyclic form of the monosaccharide, which usually results in an acyclic addition product. The latter may subsequently lose water to yield an acyclic condensation product or may cyclize. It may, however, be seen from the reaction sequence shown next that the same reaction products, whether cyclic or acyclic, may be formed by nucleophilic substitution on the hemiacetal function of a cyclic sugar. Nucleophilic substitution is much slower than nucleophilic addition, but in this case its contribution is enhanced by the large proportions of cyclic forms of the sugar present in the equilibrium mixture.

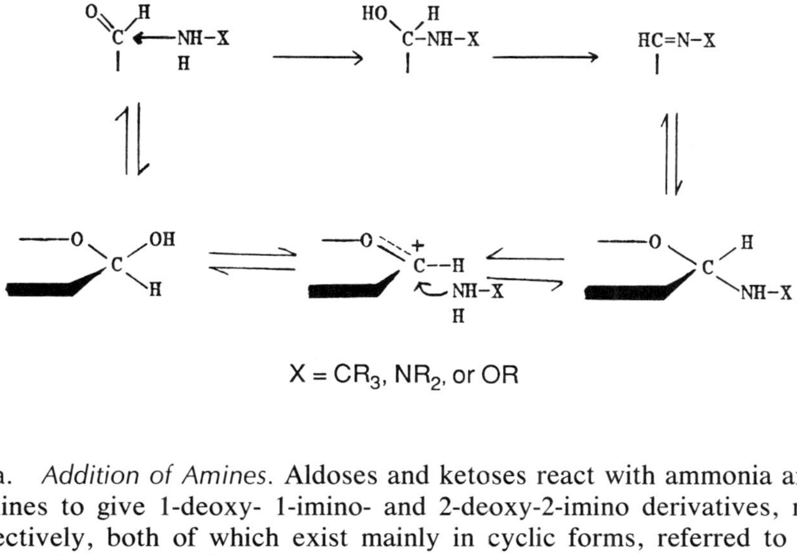

X = CR_3, NR_2, or OR

a. *Addition of Amines.* Aldoses and ketoses react with ammonia and amines to give 1-deoxy- 1-imino- and 2-deoxy-2-imino derivatives, respectively, both of which exist mainly in cyclic forms, referred to as glycosylamines. During the reaction of aldoses with amines, the 1-amino-1-deoxyglycoses, that is, the glycosylamines that are formed first, often rearrange to give 1-amino-1-deoxyketoses, called Amadori compounds. The reaction leading to their formation is referred to as the Amadori rearrangement. Unlike the glycosylamines, which exist preponderantly in cyclic forms, the Amadori compounds may be cyclic or acyclic.

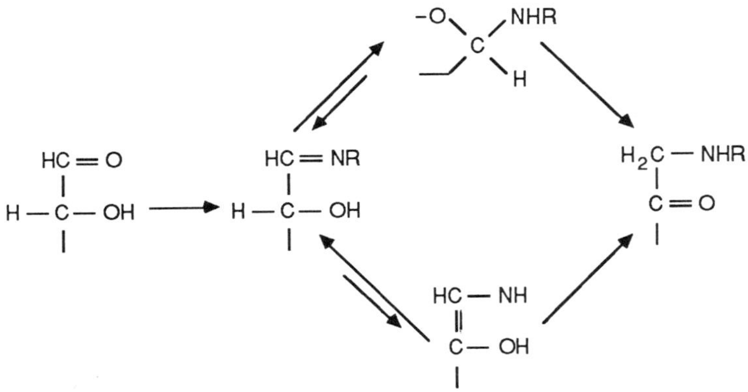

The mechanism of the reaction has been extensively studied by Weygand, who proposed the following mechanism, which involves a prototropic rearrangement of an aldimine to a 1,2-enaminol, which later ketonizes.

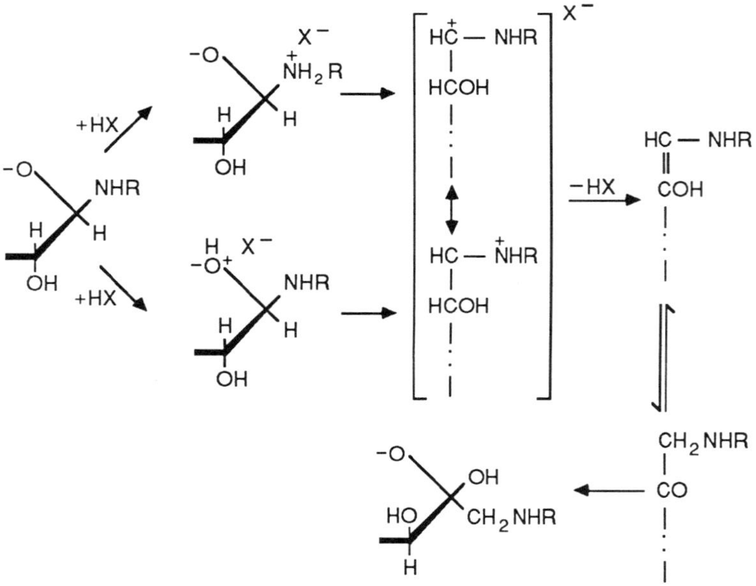

Prototropic (sigmatropic) rearrangements leading to the formation of enaminols are by no means restricted to imines. They occur during the conversion of hydrazones into osazones and are involved in the cyclization of the latter compounds, as will be seen when the addition of substituted hydrazines is discussed.

The analogous reaction between monosaccharides and amino acids or peptides is of great importance in the food industry. Its ultimate outcome is the formation of the dark polymeric products known as melanoidins, which give baked goods their characteristic color. The initial stages of this complicated reaction, collectively known as the Maillard reaction are well documented. They involve formation of *N*-glycosylamino acids, which undergo Amadori rearrangement to the ketose amino acid derivatives and then dimerize to the diketose amino acids. The latter then undergo a series of double-bond migrations, via 1,2-enolizations and 2,3-enolizations, to give mono- and dideoxyhexosuloses (see Fig. 1).

The glycosuloses then react with amino acids, and the products polymerize to give the pigments. At the same time, carbon dioxide (originating from the carboxylic group of the amino acid) and volatile components (that give baked goods their characteristic flavor) are produced. The empirical formula of the melanoidin obtained from D-glucose and glycine is $(C_{67}H_{76}N_5O_{32})_x$.

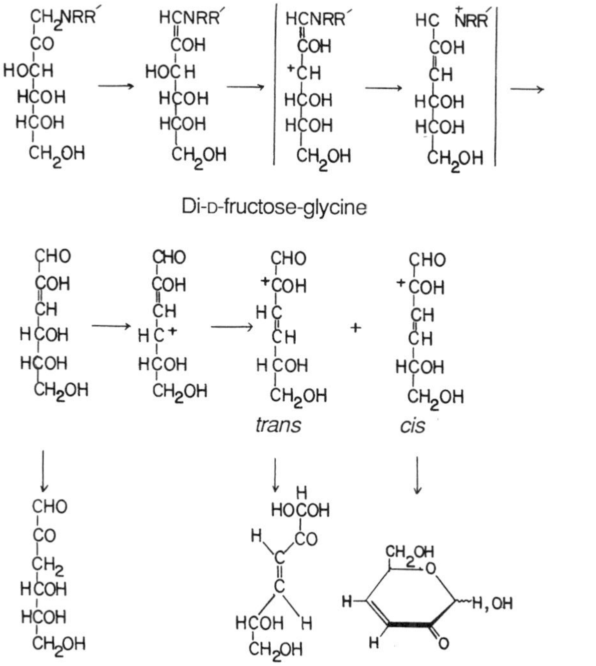

Di-D-fructose-glycine

trans cis

3-Deoxy-D-*erythro*-hexos-2-ulose

trans *cis*
3,4-Dideoxy-D-*glycero*-hex-3-enos-2-ulose

Fig. 1. Maillard reaction scheme.

b. *Condensation with Substituted and Unsubstituted Hydrazines.* The importance of the hydrazine derivatives of monosaccharides was recognized by Emil Fischer, who was the first to realize that these derivatives are useful not only for the characterization of sugars but also for the structure elucidation of monosaccharides. He used the observation that D-glucose and D-mannose yield the same osazone, which is a 1,2-bis(hydrazone), to conclude that the two monosaccharides differ only in the configuration of C-2.

The hydrazine derivatives of sugars are more reactive than the sugars from which they are prepared. When a hydrazone residue is introduced into a molecule the number of nucleophilic groups is increased by unity (the second nitrogen atom of the hydrazone acts as a strong nucleophile), whereas the number of groups capable of undergoing nucleophilic attack remains constant because nucleophiles add to the C═N group at position 1 in the same way that they do to the C═O group of the parent sugar. In the case of osazones and other bis(hydrazones), the reactivity is further enhanced because a second C═N group is introduced (in place of a less reactive H—C—OH group). As a result, the ability of osazones and bis(hydrazones) to undergo addition and cyclization reactions is greater than that of monohydrazones, which, in turn, is greater than that of the free sugars that yielded them.

In this section the hydrazine derivatives of saccharides are grouped according to the ratio of hydrazine residues combined to the sugar, starting with azines, which are produced by causing one hydrazine molecule to react with two saccharide molecules; this will be followed by a review of the chemistry of saccharide monohydrazones, where the ratio of hydrazine to sugar is 1 : 1. Finally, the chemistry of osazones and bis(hydrazones) will be treated. The latter compounds have two hydrazone residues linked to each saccharide molecule.

(i) AZINES. Aldoses and ketoses react with unsubstituted hydrazine, NH_2—NH_2, to give azines, which, on prolonged treatment with more reagent, give hydrazones. Azines are produced when one unsubstituted hydrazine molecule reacts with two sugar molecules.

$$\underset{\text{Aldose}}{\overset{|}{\mathrm{HC{=}O}}} \longrightarrow \underset{\text{Aldose azine}}{\mathrm{HC{=}N{-}N{=}CH}} \longrightarrow \underset{\text{Aldose hydrazone}}{\mathrm{HC{=}N{-}NH_2}}$$

(ii) HYDRAZONES. Saccharide hydrazones have been prepared from unsubstituted, monosubstituted, and *N,N*-disubstituted hydrazines. R. S. Tipson found that when ribose is treated with hydrazine it gives the *E*-hydrazone. The substituents attached to the hydrazine include alkyl, aryl, and heteroaryl groups as well as acyl, aroyl, thioacyl, thioaroyl, and sulfonyl groups.

Hydrazone formation is catalyzed by acids and is fastest at pH 4–5. This is why it is possible to prepare sugar hydrazones at room temperature if the substituted hydrazine salts are used in weakly acidic solutions, but, if the free hydrazine bases are used, heating becomes necessary. It should be noted that warming the former solutions causes the reaction to proceed further and yields the osazones, as will be seen in the next section.

Wolfrom and co-workers compared the rates of formation of hydrazones and observed that the reaction is fastest when an acyclic (aldehydo) sugar is used as the starting material. This suggested that the rate-determining step is either the opening of the pyranose ring or the nucleophilic substitution of the pyranose, which is, of course, more difficult than nucleophilic addition to the carbonyl group.

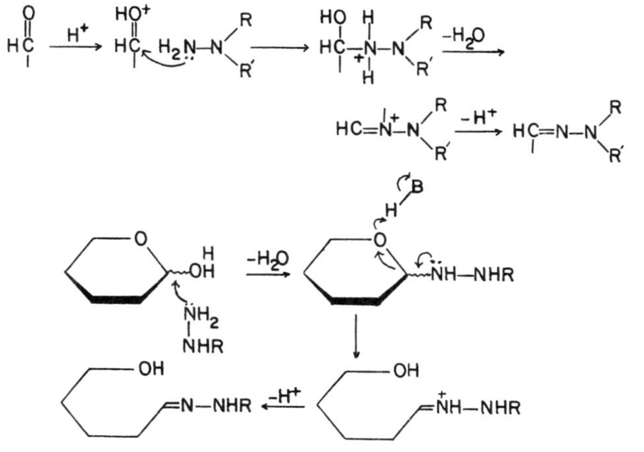

Formation of hydrazones

Sugar hydrazones can exist as equilibrium mixtures of various tautomeric forms. This is apparent from the complex mutarotation that they exhibit in solution. The most important of these structures are the acyclic Schiff bases and the two pairs of anomeric cyclic forms (the two five- and two six-membered rings). However, when crystallized, one form of the hydrazone can be isolated. For example, the crystalline form of D-galactose *N*-methyl-*N*-phenylhydrazone was shown by the following sequence of reactions to be the acyclic form.

Before the advent of chromatographic techniques, saccharide hydrazones were often prepared to purify syrupy sugars. The hydrazones formed were crystallized and the pure derivatives treated with benzaldehyde to regenerate the sugars. At present carbohydrate chemists are more interested in the reactions of hydrazones, because, as mentioned earlier, these derivatives are more reactive than the sugars from which they were prepared. An example of the reactivity of hydrazones is their ability to tautomerize. In basic media, the tautomeric azo form readily loses the imino proton and eliminates the OR group attached to C-2. This reaction was discovered by Wolfrom, who found that warming an ethanolic pyridine solution of a saccharide phenylhydrazone acetate converts it to an azoalkene. In stronger alkaline media, such as ethanolic KOH, the resulting azoalkene undergoes further rearrangements to give a three-carbon aldehyde, which undergoes nucleophilic attack by N-2 of the hydrazone residue to form phenylpyrrole.

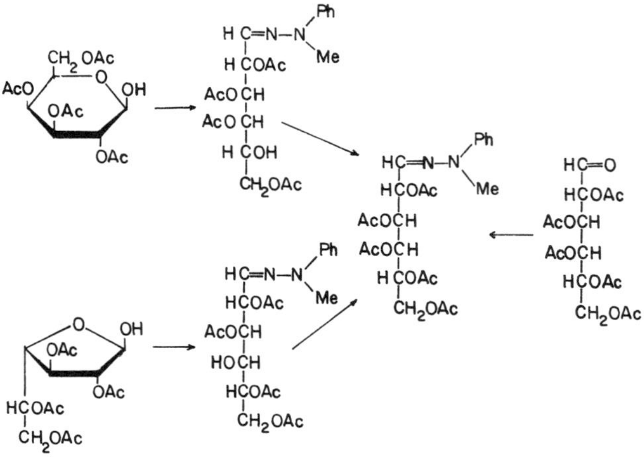

Other tautomerisms were found by Simon to occur in acidic media.

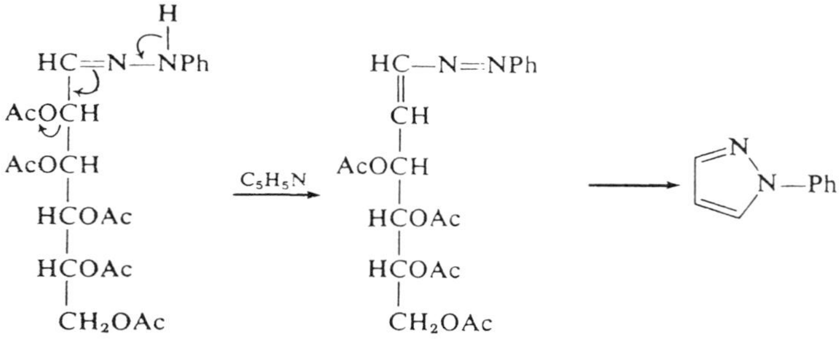

Mester and co-workers found that diazotized anilines couple with aldose monoarylhydrazones to yield red dyes known as formazans. This reaction can be used to detect imino groups of the Schiff base type

(HC=N—NH—Ar) in a hydrazone and to confirm the existence of the acyclic form. Because formazans exist as chelated resonance hybrids, the same compound may be obtained from a phenylhydrazone and a diazotized arylamine, or from an arylhydrazone and diazotized aniline.

Br—C₆H₄—NH—N=CH—HCOH—HOCH—HOCH—HCOH—CH_2OH $\xrightarrow{PhN_2^{\oplus}}$ Br—C₆H₄—N—N=C(—N=N—Ph)···H (chelate)—HCOH—HOCH—HOCH—HCOH—CH_2OH

⇅

PhNH—N=CH—HCOH—HOCH—HOCH—HCOH—CH_2OH $\xrightarrow{Br—C_6H_4—N_2^{\oplus}}$ Br—C₆H₄—N=N—C(=N—NH—Ph)···H (chelate)—HCOH—HOCH—HOCH—HCOH—CH_2OH

(iii) OSAZONES AND OTHER BIS(HYDRAZONES). Osazones and bis(hydrazones) are hydrazine derivatives of aldosuloses and diuloses, respectively. If the two hydrazone residues are attached to C-1 and C-2 of a saccharide, the derivative is referred to as an osazone and is named by suffixing "osazone" to the name of the ketose having the same carbon chain and the same configuration (irrespective of the sugar used in its preparation); for example, D-*arabino*-2-hexulose phenylosazone is the name given to the osazone obtained from D-glucose and phenylhydrazine. If the two hydrazone residues are attached at other positions, the compound is called a bis(hydrazone) and is named after the diulose by adding bis followed (between parentheses) by the name of the substituted hydrazone—for example, L-*threo*-2,4-hexodiulosonolactone 2,4-bis(phenylhydrazone).

D-*arabino*-2-Hexulose phenylosazone

L-*threo*-2,3-Hexodiulosono-1,4-lactone bis(phenylhydrazone)

The mechanism of formation of osazones and bis(hydrazones) from the corresponding dicarbonyl compounds (glycosuloses and diuloses) is straightforward. However, the mechanism of formation of osazones from aldoses and ketoses, by heating with arylhydrazine acetates, is much more difficult to explain. This is because the formation of a glycosulose derivative from an aldose or a ketose must at some stage involve the conversion of a hydroxyl group into a carbonyl group. This is usually regarded as an oxidation, and it is difficult to explain why such a strong reducing agent as a substituted hydrazine would oxidize a hydroxyl group. To overcome this problem, Weygand proposed two mechanisms that involve Amadori rearrangements instead of oxidations. The schemes start with the saccharide monoarylhydrazone, which is usually regarded as the first intermediate in osazone formation. The reactions shown are depicted for a two-carbon system, to obviate the need to have a top and a bottom hydrazone residue that may undergo a certain reaction.

Mechanisms proposed by F. Weygand

Other mechanisms have been developed that start with a transamination step initiated by the aniline produced by decomposition of phenylhydrazine. This amine can attack either the saccharide monohydrazone or one of the subsequent intermediates.

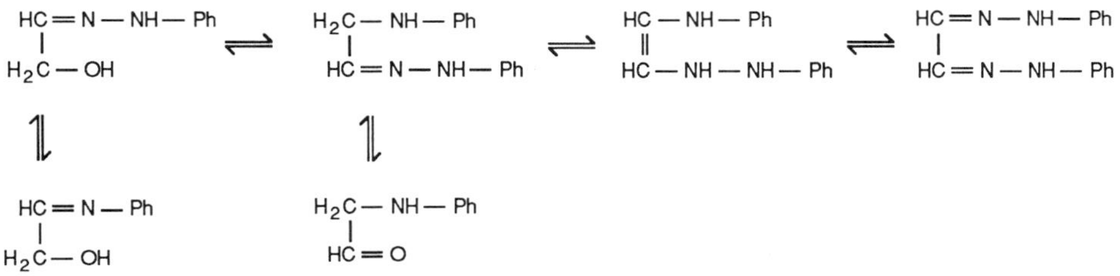

HC=N—NH—Ph | H_2C—OH ⇌ H_2C—NH—Ph | HC=N—NH—Ph ⇌ HC—NH—Ph ‖ HC—NH—NH—Ph ⇌ HC=N—NH—Ph | HC=N—NH—Ph

HC=N—Ph | H_2C—OH ⇌ (first structure); H_2C—NH—Ph | HC=O ⇌ (second structure)

(iv) REACTIONS OF OSAZONES AND BIS(HYDRAZONES). Tautomerism: Like hydrazones, which are capable of tautomerizing from the hydrazone form to both azo and hydrazino forms, osazones and bis(hydrazones) can undergo similar rearrangements. Further tautomerism may result in breaking the saccharide chain into two- and four-carbon units or two three-carbon units from a hexose derivative. [For a review see H. Simon and A. Kraus, *ACS Symp. Ser. No.* 39, 188 (1976).]

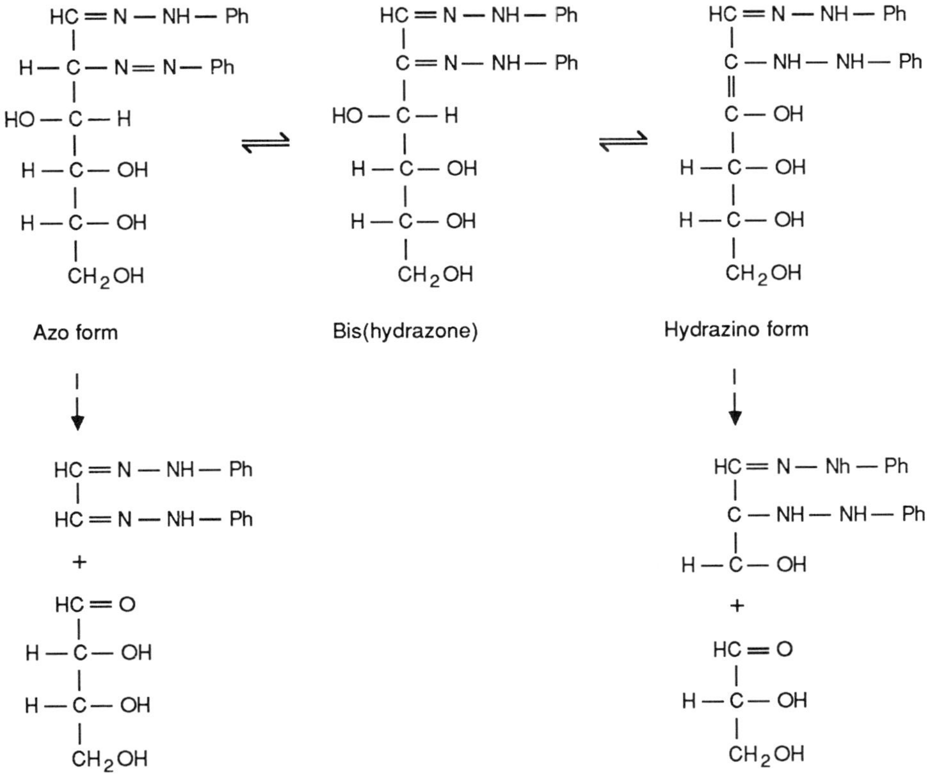

HC=N—NH—Ph | H—C—N=N—Ph | HO—C—H | H—C—OH | H—C—OH | CH_2OH

Azo form

⇌

HC=N—NH—Ph | C=N—NH—Ph | HO—C—H | H—C—OH | H—C—OH | CH_2OH

Bis(hydrazone)

⇌

HC=N—NH—Ph | C—NH—NH—Ph ‖ C—OH | H—C—OH | H—C—OH | CH_2OH

Hydrazino form

Azo form ↓

HC=N—NH—Ph | HC=N—NH—Ph

\+

HC=O | H—C—OH | H—C—OH | CH_2OH

Hydrazino form ↓

HC=N—Nh—Ph | C—NH—NH—Ph | H—C—OH

\+

HC=O | H—C—OH | CH_2OH

Chelation: The bis(hydrazone) residue of osazones exists mainly in a chelated form. This was predicted on theoretical grounds by Fieser and Fieser and used to explain why osazone formation does not proceed beyond C-2 (*N,N*-disubstituted hydrazines react with saccharides to give polyhydrazones). The chelation of osazones was later confirmed by ^{1}H-NMR and ^{15}N-NMR spectroscopy. The ^{1}H-NMR spectra of osazones revealed that the imino protons of the two hydrazone residues have widely different chemical shifts. The chelated imino protons of the hydrazone residue on C-2 are much more deshielded (δ 12–14) than those of the hydrazone on C-1 (δ 8–10). Similarly, Mester found that the ^{15}N chemical shifts of the two imino nitrogen atoms are quite different, with that of the hydrazone on C-1 occurring at a much higher field.

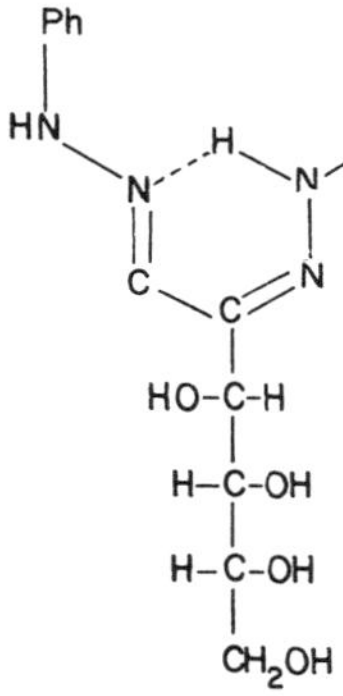

Ph
HN
H
Ph
N
N
C
C
N
HO-C-H
H-C-OH
H-C-OH
CH_2OH

Chelated form of osazone

Intramolecular Cyclizations: It was mentioned earlier that saccharide osazones and bis(hydrazones) are more reactive than the corresponding monohydrazones because they have more groups capable of undergoing nucleophilic attack (two C=N groups instead of one) and an additional nucleophile (two nitrogen atoms replacing the OH group on C-2). It should be noted that the introduction of additional carbonyl groups into an osazone or a bis(hydrazone) molecule, whether in the hydrazone moiety (by using acylhydrazines) or in the sugar moiety [as in dehydro-L-ascorbic acid bis(hydrazones)], increases the reactivity of these compounds even more. The ability of hydrazones, osazones, and bis(hydrazones) to undergo nucleophilic intramolecular cyclizations may also be enhanced by acetylation because acetyl groups are better leaving groups than OH groups.

Many intramolecular reactions exhibited by osazones and bis(hydrazones) occur via highly reactive azoalkene intermediates. Like the monohydrazones, which readily lose their imino protons and the OH group attached to the carbon atom adjacent to the hydrazone residue, saccharide osazones and bis(hydrazones) afford cyclic anhydro derivatives via similar highly reactive azoalkene derivatives. The reaction is catalyzed by either acids or bases and is greatly facilitated by acetylation of the OH leaving group.

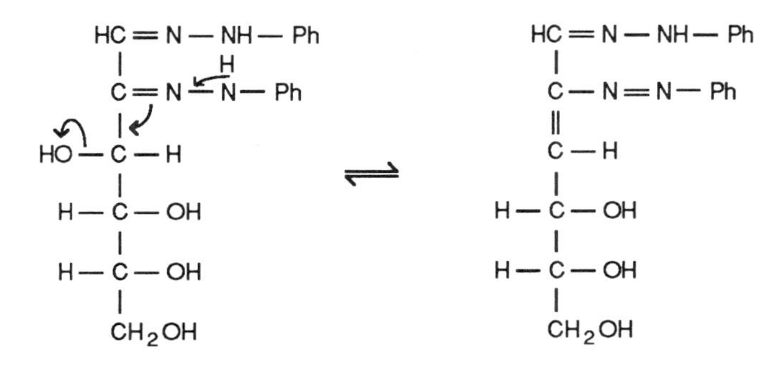

Various azoalkene derivatives have been found to be intermediates in anhydro-osazone formation, as seen in the three reactions described next.

1. Diels monoanhydro-osazone. This derivative is prepared in acidic medium by refluxing the osazone with either methanolic HCl or a mixture of acetic acid and acetic anhydride (in which case the acetate is formed). The reaction was discovered by Diels (of Diels–Alder reaction fame), and for this reason the products obtained are referred to as Diels anhydro-osazones. The mechanism of the reaction was elucidated by Simon, who recognized the importance of the azoalkene intermediate, which is attacked by the O-6 atom to form a 3,6-anhydro ring. Of the two anomers formed, the one having the bis(hydrazone) residues and the OR on C-4 *trans* usually preponderates (attack from the less hindered side).

Formation of Diels anhydro-osazone

2. Percival's dianhydroosazone. This product is obtained in basic media by treating osazone acetates with ethanolic KOH. The acetylated azoalkene first formed undergoes further migration of the double bonds; the intermediate thus formed undergoes nucleophilic attack by the nitrogen atom of the hydrazone residue on C-1 to form a diazine ring, which cyclizes further to form a 3,6-anhydro ring. Because of the rigidity of the fused rings of this dianhydroosazone, it exists in only two forms, one obtained from hexoses of the D series and the other from hexoses of the L series.

Azoenehydrazone

Percival osazone

3. Dianhydroosazones of the pyrazole type. The third type of anhydroosazone was discovered by the author, who found that refluxing saccharide osazones with a mixture of acetic anhydride and acetic acid yields hydrazones of hydroxyalkylpyrazoles. As in the previous reactions, an azoalkene intermediate undergoes migration of double bonds to give a 3,4-enol hydrazone. Because in this case the nitrogen atom of the hydrazone residue on C-1 is deactivated by an acetyl group introduced during the refluxing with acetic anhydride, nucleophilic attack occurs via the nitrogen atom of the 2-hydrazone group, to form the pyrazole ring. The insufficient length of the carbon chain prevents formation of a 3,6-anhydro ring.

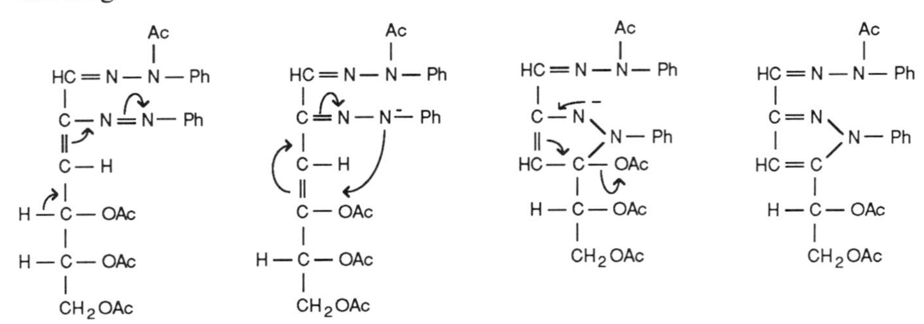

Azoenehydrazone

Pyrazole

Another hydrazo intermediate is probably involved in the oxidative aromatization of osazones to osotriazoles (1,2,3-triazoles). This reaction was discovered by Hudson, who found that, when treated with Cu^{2+}, sugar osazones yield the corresponding triazoles. By using radioactive tracers, Weygand later found that during this conversion most of the aniline liberated originated from the hydrazone residue attached to C-1. It seems that chelation of the osazone renders the transfer of the imino proton from the nitrogen atom of the hydrazone residue on C-2 to that of the residue on C-1 by a 1,5 sigmatropic shift. Subsequent loss of $PhNH^-$ results in the protonated triazole. Because this base is very weak, it loses its proton at neutral pH.

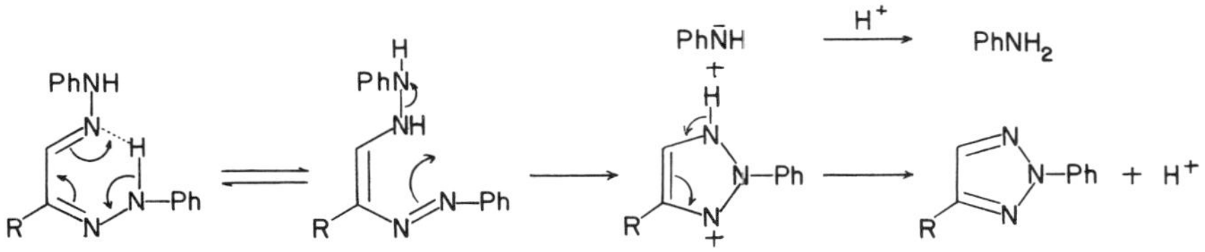

Formation of osotriazoles

The nucleophilicity quotient (NQ) is a measure of the susceptibility of polyfunctional molecules to cyclization by intramolecular nucleophilic addition. Sugars and sugar derivatives possess more nucleophilic species than nucleophile acceptors (for example, there are five nucleophilic oxygen atoms versus one carbonyl group in the acyclic form of an aldohexose). Accordingly, the number of nucleophile acceptors has a greater (and sometimes controlling) influence on the ability of the molecule to cyclize than the nucleophilicity of the attacking groups. By using integers to designate the number of nucleophile acceptors in the molecule (for example, C=O, C=NR, C=C—C=O, and C=C—C=NR) and fractions to designate the nucleophilicity (n) of the attacking groups (for example, OH or NH groups), one can get a numerical value that reflects the ability of a compound to undergo nucleophilic reactions.* For example, free aldoses have a nucleophilicity quotient of 1.4, which signifies that they possess 1 nucleophile acceptor (C-1 of the acyclic form) and that the nucleophilicity of the attacking oxygen atom (as determined from tables) is 4. Saccharide hydrazones and oximes have a slightly higher NQ value (1.6), and saccharide osazones have an NQ of 2.6, which reflects their reactivity. Compounds having an additional carbonyl group (in the saccharide moiety or in the hydrazone residue) are even more reactive. For example, L-ascor-

* The value of n is the affinity of the nucleophile toward an acceptor; it can be obtained from tables, or from kinetic measurements using a Hammett-like equation ($\log k/k_0 = ns$, where s is the sensitivity of the nucleophile acceptor to the nucleophile).

bic acid bis(phenylhydrazones) and glycosulose bis(benzoylhydrazones) have an NQ value of 3.6.

c. *Addition of Hydroxylamine.* Reducing sugars react with hydroxylamine to give oximes, which, like the hydrazones, may be acyclic or cyclic. Owing to their high solubility in water and ethanol, saccharide oximes, unlike their hydrazone counterparts, have not been extensively used for the characterization of sugars. On the other hand, they have been prepared for use in the Wohl degradation (see p. 106) for descending the series. This reaction makes use of the fact that, on refluxing with acetic anhydride, the oxime readily loses water (or acetic acid, if it is first *N*-acetylated) and is converted into the nitrile. Then, using the fact that the addition of HCN to aldehydes is reversible, the nitrile formed is hydrolyzed to an aldose having one carbon atom less than the starting oxime. If carried out with ammonia, the hydrolysis produces an undesired intermediate bis(acetamido) derivative, which may be hydrolyzed with acid before the desired aldose can be liberated.

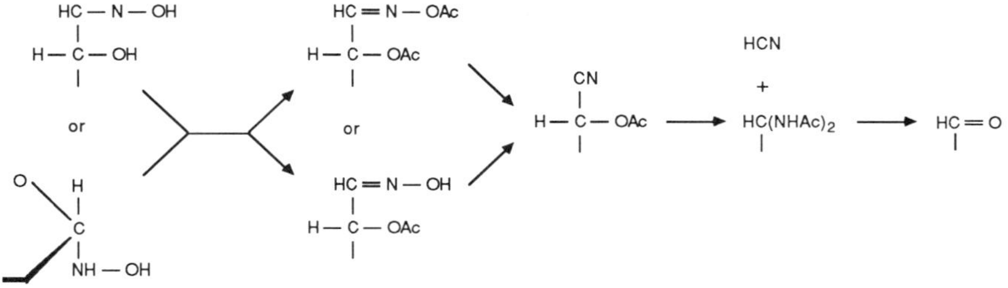

The facile elimination of acetic acid from oximes on boiling with acetic anhydride has been used to prepare some triazoles from oxime hydrazones. Because on oxidation the bis(hydrazone) of dehydro-L-ascorbic acid gives a bicyclic derivative instead of a triazole, the latter was prepared by dehydration of the hydrazone oxime with boiling acetic anhydride.

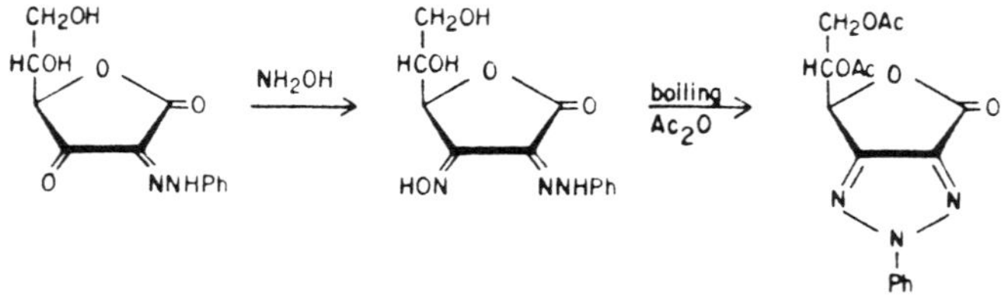

Like the saccharide hydrazones, the saccharide oximes yield 1-amino-1-deoxyalditols on reduction.

3. Addition of Oxygen and Sulfur Nucleophiles

This section discusses the addition of one or two molecules of the nucleophiles HO^-, HS^-, RO^-, and RS^- to the carbonyl group of saccharides.

a. *Addition of HO^- and HS^-.* Addition of 1 mole of these two nucleophiles results in the reversible formation of reaction products that are detectable in solution by NMR spectroscopy but not isolated in the free forms. This is not surprising, because their formation is constantly competing with intramolecular nucleophilic attack by one of the OH groups (present in the sugar molecule) that is in close proximity to the carbonyl group, to form one of the cyclic forms of the sugar.

$$\mathrm{HO-\overset{H}{\underset{|}{\overset{|}{C}}}-OH} \rightleftharpoons \mathrm{\overset{H}{\underset{|}{\overset{|}{C}}}=O} \rightleftharpoons \mathrm{HS-\overset{H}{\underset{|}{\overset{|}{C}}}-OH}$$

b. *Addition of RO^- and RS^- Nucleophiles.* Addition of 1 mole of these nucleophiles to the carbonyl group of a saccharide yields the hemiacetal or thiohemiacetal, which, in solution, remains in equilibrium with the free saccharide. Addition of 2 moles of the nucleophile yields stable acetals or dithioacetals.

If one of the OH groups of the acyclic form of the sugar is the nucleophile, the product is one of the four possible cyclic structures of a sugar (α- and β-furanose and α- and β-pyranose). If, instead, the nucleophile originates from without the sugar molecule, hemiacetals or thiohemiacetals are obtained, depending on the nucleophile. Because these compounds are constantly in equilibrium with the acyclic form, they are not isolated and they have little importance in preparative chemistry.

$$\mathrm{RO-\overset{H}{\underset{|}{\overset{|}{C}}}-OH} \longleftarrow \mathrm{\overset{H}{\underset{|}{\overset{|}{C}}}=O} \longrightarrow \mathrm{RS-\overset{H}{\underset{|}{\overset{|}{C}}}-OH}$$

Much more important are the acetals and dithioacetals, which are obtained by reaction of 1 mole of monosaccharide with either 2 moles of the same nucleophile or with 1 mole of each of two different ones. Three types of acetal are possible, depending on the source of the nucleophile(s) involved in the reaction: (a) if both nucleophile molecules do not arise from the sugar molecule, the product is an acyclic acetal or dithioacetal; (b) if one nucleophile forms part of the sugar molecule and one is not a part of it, a cyclic acetal or thioacetal is obtained which is referred to as a glycoside or 1-thioglycoside; (c) if both nucleophile molecules are part of the same sugar molecule, the product is a bicyclic acetal. Acetals of types (b) and (c) are, in reality, derivatives of the cyclic forms of sugars, and as such they do not belong here among the reactions of the carbonyl group. Their discussion will accordingly be deferred to the next section.

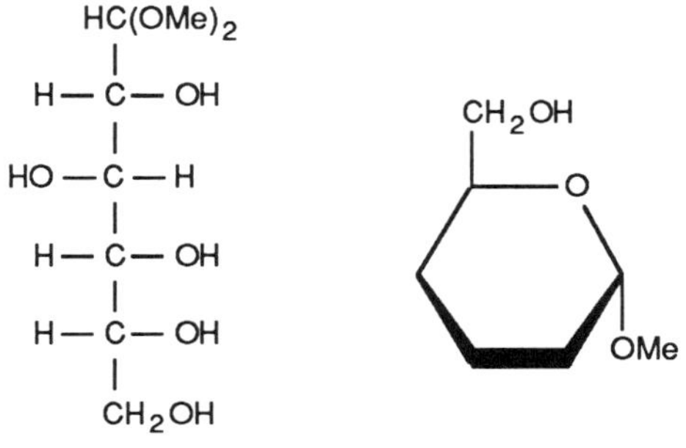

$HC(OMe)_2$
H—C—OH
HO—C—H
H—C—OH
H—C—OH
CH_2OH

(a)

CH_2OH, O, OMe

(b)

CH_2—O, O

(c)

In general, the best route for preparing the acyclic forms of sugars and acyclic sugar derivatives is from the dithioacetals; another way is to prepare an acyclic hydrazone of the sugar, acetylate it, and then remove the hydrazone residue from the acetate with benzaldehyde. Because of the importance of dithioacetals in synthesis, their preparation and their conversion into acetals and other useful derivatives will be examined. Dithioacetals are readily obtained in acidic media by treating monosaccharides with the desired thiol, usually ethanethiol, in the presence of concentrated HCl. Dithioacetals are easier to form than acetals because sulfur is less basic than oxygen; accordingly, the population of unreactive protonated oxonium ions is larger than that of protonated sulfonium ions.

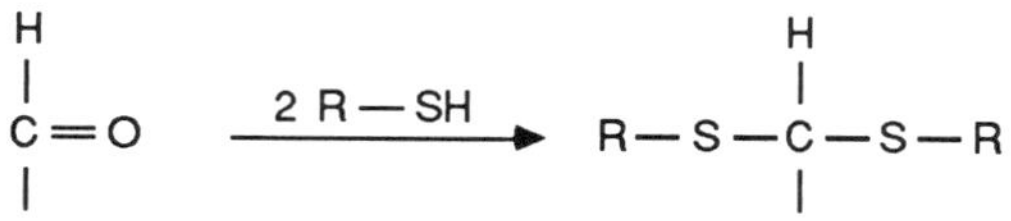

$$\mathrm{H{-}C{=}O} \xrightarrow{2\ \mathrm{R{-}SH}} \mathrm{R{-}S{-}CH{-}S{-}R}$$

The demercaptalation (removal of the thioacetal residues) is usually conducted in the presence of a mixture of yellow mercuric oxide and cadmium carbonate or mercury chloride. If hydrolysis is needed in order to prepare the aldehydo sugar, water is added to the mixture, making sure the reaction is performed on the peracetylated dithioacetal (to prevent formation of the cyclic forms).

HC=O / H—C—OH / HO—C—H / H—C—OH / H—C—OH / CH_2OH $\xrightarrow[H^+]{EtSH}$ $HC(SEt)_2$ / H—C—OH / HO—C—H / H—C—OH / H—C—OH / CH_2OH $\xrightarrow{Ac_2O}$ $HC(SEt)_2$ / H—C—OAc / AcO—C—H / H—C—OAc / H—C—OAc / CH_2OAc $\xrightarrow[H_2O]{HgO}$ HC=O / H—C—OAc / AcO—C—H / H—C—OAc / H—C—OAc / CH_2OAc

If an alcohol (for example, methanol) is used in this reaction instead of water, the dimethyl acetal is obtained.

$$HC{=}O \xrightarrow{EtSH} HC(SEt)_2 \xrightarrow{MeOH} HC(OMe)_2$$

Because dialkyl diacetals are extremely sensitive to acid, large proportions of cadium carbonate are needed; if no carbonate is used, a mixture of the dimethyl acetal, the two methyl furanosides, and the two methyl pyranosides is obtained. To increase the yield of the kinetically favored methyl furanosides, this reaction is carried out at room temperature.

CH_3, HO, NHAc, OH ⟶ H / EtSCSEt / CH_2 / HCNHAc / HCOR / HCOR / CH_3 ⟶ CH_3 / HOCH, NHAc, OMe + CH_3 / HOCH, OMe, NHAc

On treatment with bromine, acetals and thioacetals readily undergo nucleophilic substitution at C-1 to afford bromides. These intermediates have been used by Wolfrom and by Horton to prepare acyclic nucleosides.

H / (EtS—C—SEt) / AcO—C—H / AcO—C—H / H—C—OAc / CH_2OAc ⟶ H / (Br—C—SEt) / AcO—C—H / AcO—C—H / H—C—OAc / CH_2OAc ⟶ H / (RO—C—SEt) / AcO—C—H / AcO—C—H / H—C—OAc / CH_2OAc ⟶ H / (RO—C—Br) / AcO—C—H / AcO—C—H / H—C—OAc / CH_2OAc (+ NH_2-purine–N–HgX) ⟶ (adenin-9-yl) / (RO—C—H) / AcO—C—H / AcO—C—H / H—C—OAc / CH_2OAc

B. Nucleophilic Substitution Reactions of the Anomeric Carbon Atom

The anomeric carbon atom is the most reactive carbon atom in furanoses, pyranoses, and their derivatives. Although it is less susceptible to nucleophilic attack than the carbonyl group of acyclic sugars, it is significantly more reactive than the remaining hydroxyl-bearing carbon atoms of the saccharide. The affinity of the anomeric carbon atom toward nucleophiles depends on the nature of the leaving group. Resonance-stabilized carbonium ions can result, which undergo direct reactions via S_N1 mechanisms or synchronous reactions by S_N2-type mechanisms. The rates of these reaction will depend on the nucleophilicity of the attacking group and the nature of the leaving group (how good a leaving group it is). The configuration of the product of a nucleophilic displacement at the anomeric center in a cyclic sugar is influenced by the groups adjacent to the anomeric carbon, specifically the ring oxygen and the substituent on C-2. By virtue of the formation of cyclic oxonium ions, the formation of axially oriented anomers may be favored. This is due to a stereoselectively favored axial attack which leads to the $^4C_1$ form directly, as well as to the anomeric effect discussed earlier.

The groups attached to C-2 can be classified according to whether they direct the entering group as (1) participating groups, which comprise esters such acetates and benzoates and yield the trans anomer with respect to the C-2 substituent, and (b) nonparticipating groups, such as ethers (benzyl groups) and deoxy functions, which favor the axial anomers.

Participating groups such as esters can form with the anomeric carbon positively charged five-membered ring intermediates (dioxolenium ions), which force the nucleophile to attack C-1 from the side opposite the ring. The resulting thermodynamic products will have the substituent on C-1 in a trans configuration with respect to the one on C-2, even if the former is equatorially oriented and not favored by the anomeric effect. For example, penta-O-acetyl-D-glucopyranose will preferentially yield the equatorially oriented β anomer, and penta-O-acetyl-D-mannopyranose the axially oriented α anomer.

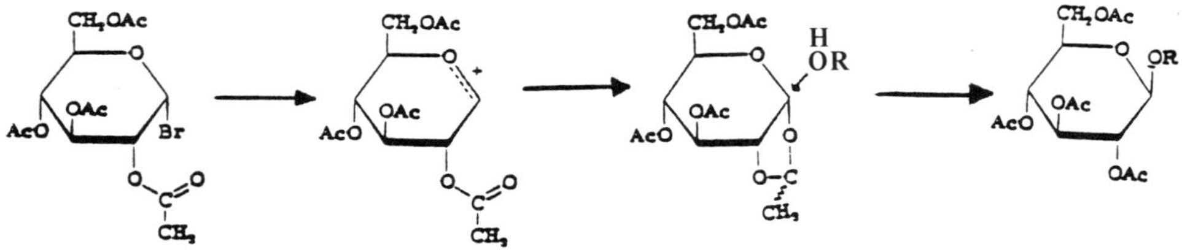

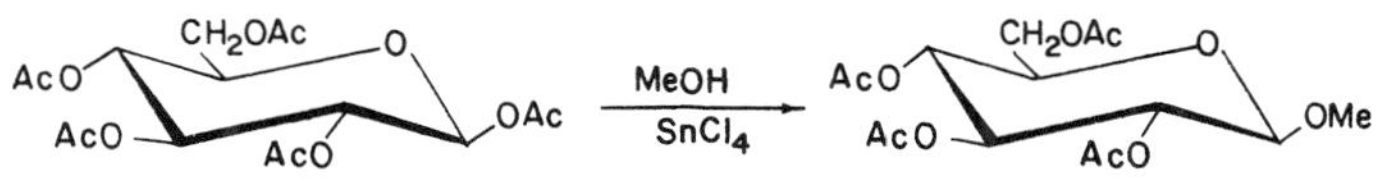

1. Participating Groups (a Means of Access to 1,2-trans Anomers)

Ester groups attached to position 2 of glycopyranosyl derivatives having good leaving groups at the anomeric position readily form cyclic resonance-stabilized acyloxonium ions. The RO^- group of alcohols will attack these intermediates from the less hindered (exo) side and give glycosides having 1,2-trans arrangements. Of course, attack may also occur on C-2 of the oxonium ion ring to give an orthoester, but this readily undergoes rearrangement to give the same glycoside. The attack in this last case is likewise exo. It should be noted that a 1,2-trans arrangement will counteract the anomeric effect in the case of allose, glucose, gulose, and galactose and lead to β-D anomers in the $^4C_1$ conformation, whereas in the case of altrose, mannose, idose, and talose the tendency to form the 1,2-trans anomer will reinforce the anomeric effect and yield the α-D anomer.

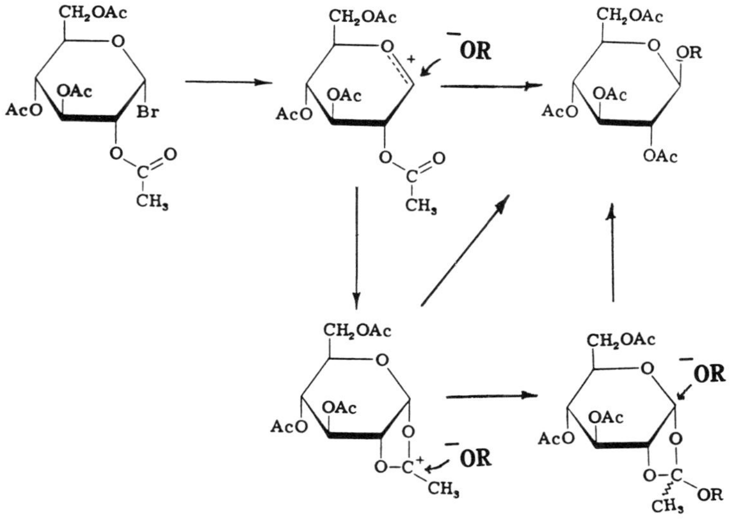

Nonparticipating groups such as ethers, cyclic acetals, or deoxy functions attached to C-2 are unable to form such acyloxonium ions, and the orientation of the resulting derivative is governed solely by the requirements of the anomeric effect (the α-D anomers being preferentially formed).

2. Displacements of Hemiacetal OH Groups and Anomeric Ester Groups

The most important displacement reactions of the hemiacetal OH groups and their esters are their displacements by OR groups to form glycosides and by X groups to form glycosyl halides.

a. *Displacements by OR groups (Glycosidation).* The anomeric hydroxyl groups of cyclic sugars and their esters can be exchanged in acid media by the OR group of alcohols and phenols. The displacement of the hemiacetal hydroxyl group by an OR group is known as the Fischer method of glycoside formation, and the similar exchange of an anomeric ester group is referred to as the Helferich method.

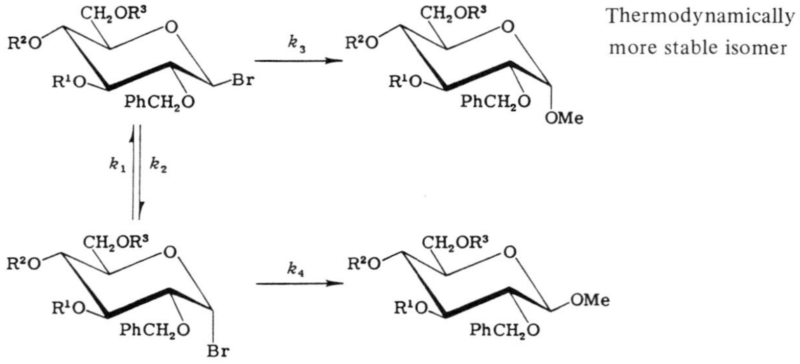

Thermodynamically more stable isomer

(i) FISCHER GLYCOSIDATION METHOD. An example of the Fischer glycosidation method is the conversion of D-glucose into two methyl D-glucofuranosides and two methyl D-glucopyranosides by treatment with methanolic HCl. The first products isolated are the two furanosides, methyl α-D-glucofuranoside and methyl β-D-glucofuranoside. The furanosides are kinetically favored because closure of five-membered rings is faster than that of six-membered rings. If the reaction time is prolonged or if reflux temperatures are used, the thermodynamically favored pyranosides will preponderate and crystalline methyl α-D-glucopyranoside and methyl β-D-glucopyranoside will be obtained in high yields.

β-D-glucose

MeOH, 0.7% HCl, 20° C → α, β Methyl D-glucofuranosides

MeOH, 4% HCl, reflux temp. → α, β Methyl D-glucopyranosides

The mechanisms of glycoside formation for furanosides and pyranosides are quite similar, and so the discussion will be restricted to one type, namely the formation of furanosides. Figure 2 depicts the formation of a methyl α-D-glycofuranoside from an equilibrium mixture of an acyclic D-aldopentose and the β-D-aldofuranose. The first step of this reaction, as in all glycosidations, consists in the protonation of the sugar (in this case the cyclic and acyclic forms), and it may be seen that four ions will be formed, either directly or after loss of water; two of these ions have positive charges on oxygen atoms and two on carbon atoms. The former two are the more stable, because each atom in the molecule (except hydrogen) is surrounded by an octet. Attack by the methanol oxygen atom on the more stable ions (having charges on oxygen) can occur by synchronous S_N2 mechanisms to yield methyl α-D-aldofuranoside or an acyclic hemiacetal, which can cyclize to the methyl α-D-aldofuranoside or its β-D anomer. The same products are formed by an S_N1 reaction initiated by attack of the methanol oxygen atom on the carbonium ions (having the charges on carbon).

The reactions involving the conversion of the kinetically favored aldofuranosides into the thermodynamically favored aldopyranosides are complex and include anomerization of the furanosides, formation of the pyranosides, and anomerization of the latter. Table II shows the composition at equilibrium of the pentoses of the D series.

A study of the products of the Fischer reaction reveals the following points. (a) At equilibrium, the pyranose forms usually preponderate, as the six-membered rings are the thermodynamically favored forms of glycosides. (b) As a rule, aldoses having adjacent bulky groups (OH or OMe groups) in the cis orientation will give less of the furanosides on equilibration, because these substituents are closer together (and repel one another

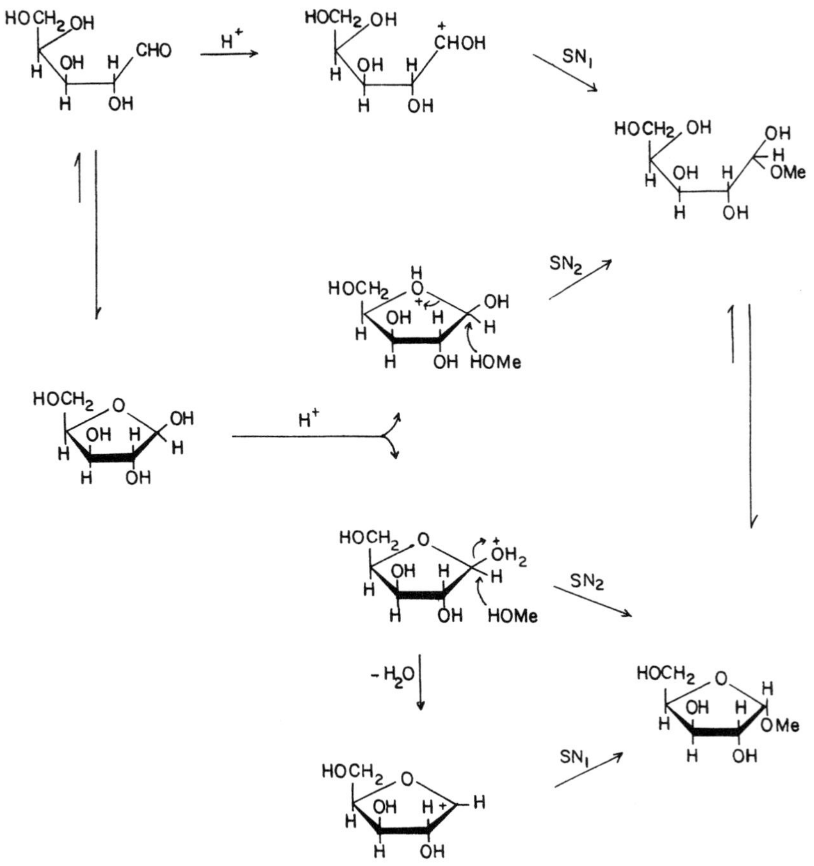

Fig. 2. Scheme for D-glucofuranoside formation.

TABLE II
Percentage of Methyl Aldosides at Equilibrium[a]

D-Aldopentose	Furanoses α	Furanoses β	Pyranoses α	Pyranoses β
Ribose	5.2	17.4	11.6	65.8
Arabinose	21.6	6.8	24.5	47.2
Xylose	1.9	3.2	65.1	29.8
Lyxose	1.4	—	88.3	10.3

[a] From a D-aldopentose and 1% methanolic HCl at 35°C.

more) in furanosides than in pyranosides. For the same reason, a trans relationship of adjacent bulky groups is more favorable in furanosides than in pyranosides. This is why furanosides having C-1 and C-2 trans (for example, the α anomer of D-arabinose and D-lyxose and the β anomer of D-ribose and D-xylose) are more stable and exceed the concentration of their anomers. It may also be seen from Table II that the all-trans methyl α-D-arabinofuranoside is present in high concentration in the equilibrium mixture. In the case of the methylated derivatives of D-arabinose (which have bulkier OMe groups), the all-trans orientation is so favorable in the five-membered ring that the concentration of the furanosides exceeds that of the pyranosides at equilibrium. (c) Because of the anomeric effect, the anomer having an axial methoxyl group always preponderates between the two pyranosides (in furanosides no substituent exists in a truly axial position). (d) Although this fact is not reported in the table, it must be remembered that, during the formation of methyl glycosides by the Fischer method, a small proportion of the acyclic acetal is always produced. This by-product is formed by nucleophilic addition to the carbonyl group of the acyclic sugar as well as by nucleophilic substitution on the furanose form of the sugar, which can cause the ring to open.

(ii) HELFERICH METHOD OF GLYCOSIDE FORMATION. In the Helferich method of glycoside formation a peracetylated sugar (obtained by treating a free sugar with acetic anhydride in pyridine) is allowed to react with a phenol or an alcohol in the presence of an acid catalyst, such as $ZnCl_2$ or, better, *p*-toluenesulfonic acid. Because peracetylated sugars usually exist in the pyranoid form, this method is a convenient way in which to prepare glycopyranosides. If a peracetylated furanose is the starting compound, the product will, evidently, be a glycofuranoside. If mild conditions are used (for example, a short heating time with *p*-toluenesulfonic acid), the conversion may be achieved with retention of configuration. However, if drastic conditions are used, anomerization occurs (see later) and the anomer having a bulky group axial on C-1 will preponderate. As in the Fischer method, the intermediate here is a resonance-stabilized carbonium ion.

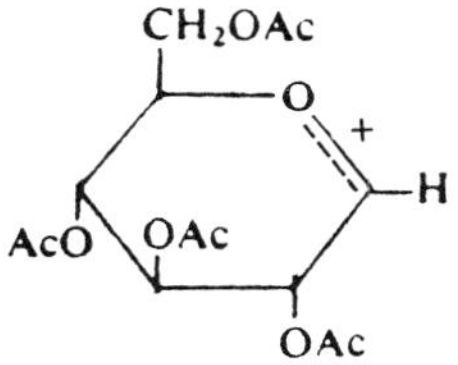

CH_2OAc
O
+
H
AcO
OAc
OAc

b. *Displacement of Anomeric Ester Groups by a Halogen.* Because halogens are good leaving groups, the glycosyl halides are important intermediates in carbohydrate syntheses. The most important derivatives are the glycosyl chlorides and bromides, the iodides usually being too reactive and therefore not stable over any length of time, and the fluorides too inert and requiring different conditions for reaction. The glycosyl halides may be prepared by bubbling HCl or HBr into a solution of a peracetylated or perbenzoylated monosaccharide. Alternatively, as will be seen later, they can be obtained in the same way from glycosides. Although many halides are isolated in crystalline form, they are often caused to react directly with a nucleophile after removal of the dissolved acid. The isomer having an axial halogen is usually preponderant (see anomeric effect).

One of the best-known reactions of glycosyl halides is the Koenigs–Knorr method of glycosidation. This method involves the reaction of a glycosyl halide with an alcohol in the presence of a heavy-metal catalyst. For example, tetra-*O*-acetyl-α-D-glucopyranosyl bromide reacts with methanol in the presence of silver carbonate (which acts as an acid acceptor) to produce methyl tetra-*O*-acetyl-β-D-glucopyranoside in high yield. The reaction mechanism is presumed to involve an intermediate carbonium ion.

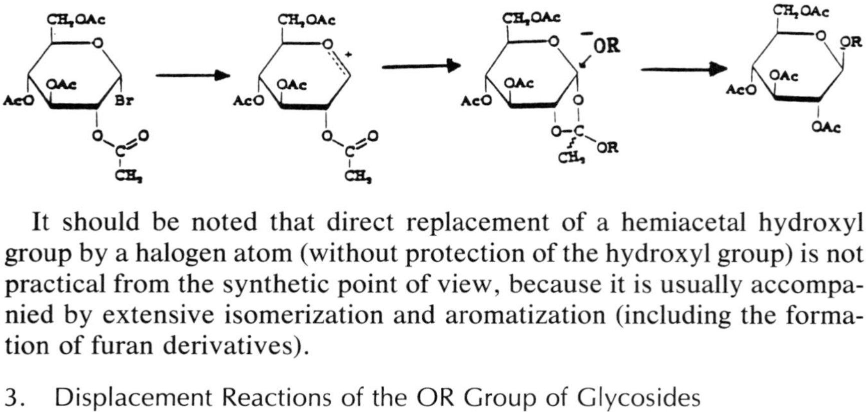

It should be noted that direct replacement of a hemiacetal hydroxyl group by a halogen atom (without protection of the hydroxyl group) is not practical from the synthetic point of view, because it is usually accompanied by extensive isomerization and aromatization (including the formation of furan derivatives).

3. Displacement Reactions of the OR Group of Glycosides

Nucleophilic substitution reactions on the anomeric carbon atom are not restricted to displacement of the anomeric hydroxyl or acetoxyl group just discussed, but may involve displacement of the OR group of glycosides by a nucleophile. The nucleophiles discussed in this section are the OH and OR groups, as well as halides. Acid catalysts are always needed in the anomeric displacement reactions in order to produce the reactive species, a resonance-stabilized carbonium ion. The latter is usually formed when the oxygen atom of the OR group attached to the anomeric carbon atom becomes protonated and the aglycon is eliminated as ROH. If the glycoside reacts with the same nucleophile found in its aglycon (for example, if a methyl glycoside is treated with methanol in the presence of an acid catalyst), anomerization will result; i.e., the two anomers of the same glycoside will be produced, and the anomer having the OR group axially attached will preponderate (the anomeric effect). If, on the other hand, the glycoside reacts with a different nucleophile, for example, if a methyl glycoside is treated with water (in a hydrolysis) or with another alcohol (in a transglycosidation), the OR group will be replaced by an OH group in the first case and by an OR′ group in the second. It should be noted that, in such reactions, if the hydroxyl group attached to C-4 of glycopyranosides and C-5 of glycofuranosides is not blocked, the foregoing displacement reactions will give mixtures of α- and β-furanoses (or furanosides) and α- and β-pyranoses (or pyranosides), irrespective of whether the starting glycoside is a furanoside or a pyranoside.

Another important nucleophile that can react with glycosides is X^-, to produce glycosyl halides, which are synthetically valuable intermediates.

a. *Hydrolysis of Glyosides (OR → OH).* In general, glycosides are hydrolyzed by acids and are stable toward bases. Exceptions to this rule are glycosides having, as the aglycon, a group derived from a phenol, an enol, or an alcohol with an electronegative group in the β position, as these are labile toward bases.

Aldofuranosides are hydrolyzed much faster than (at 50–200 times the rate of) the thermodynamically more stable aldopyranosides, and among the furanosides the less stable isomers (those having adjacent bulky groups in the cis orientation) are hydrolyzed fastest. Similarly, methyl β-D-glycopyranosides are hydrolyzed faster than their (more stable) α-D anomers because of the anomeric effect (the situation is reversed when bulky aglycons are used, because of the considerable axial–axial interactions). The ketofuranosides and ketopyranosides are also hydrolyzed faster than the corresponding aldosides. Finally, aldopyranoses and aldofuranoses having a 2- or 3-deoxy or a 2,3-dideoxy functionality are hydrolyzed considerably faster than their hydroxylated counterparts; this is due in part to a decrease in steric interaction during the conversion of the chair conformer of the glycopyranoside into a half-chair conformer of the carbonium ion and during the biomolecular displacement of the alcohol by water in the glycofuranosides (see the mechanisms given later).

The accepted mechanism for the acid hydrolysis of glycosides starts, as usual, with protonation. Although protonation of either the glycosidic oxygen atom (that of the -OR group attached to the anomeric carbon atom) or the ring oxygen atom is possible, there is evidence that it is usually the former that is protonated. The next step in the reaction, namely the formation of a carbonium ion by shifting the electrons of the C–O–R bonds, can be achieved in two ways, either by breaking the C–O bond and shifting the positive charge to the anomeric carbon atom or by breaking the O–R bond and shifting the charge to the R group of the aglycon. The first route is favored because the cyclic oxonium ion is stabilized by resonance with the form having the charge on the ring oxygen atom, and most hydrolyses follow this route. In rare cases, for example, during the acid hydrolysis of *tert*-butyl glycosides, the bond between the oxygen atom and the R group is preferentially broken because of the remarkable stability of the *tert*-butyl carbonium ion. The foregoing mechanisms have been confirmed by conducting hydrolyses of glycosides in ^{18}O-labeled water and locating the labeled oxygen atom in the products (the free sugar or the alcohol). With most aglycons, the label was found to be attached to the free sugar liberated, whereas with *tert*-butyl glycosides the label was found to be on the alcohol formed. Some studies [see A. J. Bennett and M. L. Sinnott, *J. Am. Chem. Soc.* **108,** 7287 (1986)] suggest that the protonated transition state of methyl α-D-glucopyranoside tends to assume a boat conformation, whereas that of the β anomer is firmly held in the $^{1}C_4$ conformation.

The hydrolysis of substituted aryl D-glucosides was studied by Rydon, who found that in the β series the reaction constant k increases with the electron-releasing ability of the group (Hammett constant 0.66), whereas in the α series the changes are minimal (Hammett constant = 0.006). To explain this behavior, it was suggested that aryl β-D-glucosides, like most other glycosides, are protonated on the glycosidic oxygen atom, whereas aryl α-D-glucosides are protonated on the ring oxygen atom. More work is needed to explain why this is the case, if, indeed, it is actually the reason for the behavior observed.

b. *Anomerization (OR → OR).* Anomerization is the acid-catalyzed isomerization of a group (OR) attached to the anomeric carbon atom of a cyclic saccharide derivative. The reaction is initiated by protonation of the OR group of the glycoside and is followed by elimination of the alcohol group (ROH) to afford a carbonium ion. This then undergoes nucleophilic attack by an OR group (identical to the OR originally present in the glycoside). Of the two anomers possible, that having the bulkier axial group preponderates (the anomeric effect).

If a Lewis (aprotic) acid such as $TiCl_4$ or $SnCl_4$ is used as the catalyst, the acid (A) coordinates with the oxygen atom of the aglycon group (OR) to give an ion-pair intermediate, which loses the group AOR^- and yields the glycoside. The mechanism shown was proposed by Lemieux to explain the formation of α-D-glycopyranosides by participation of an ester group on C-2.

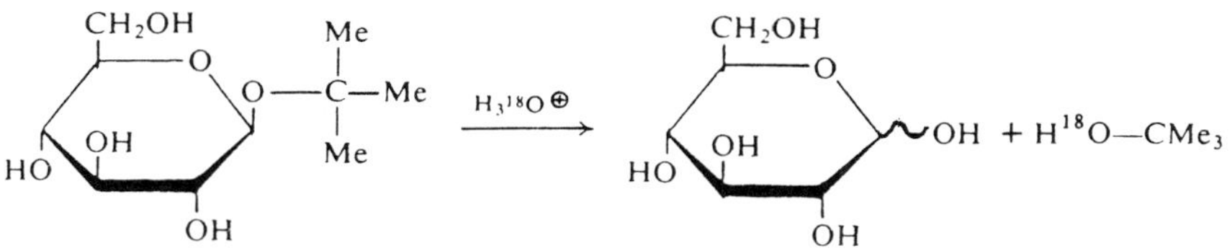

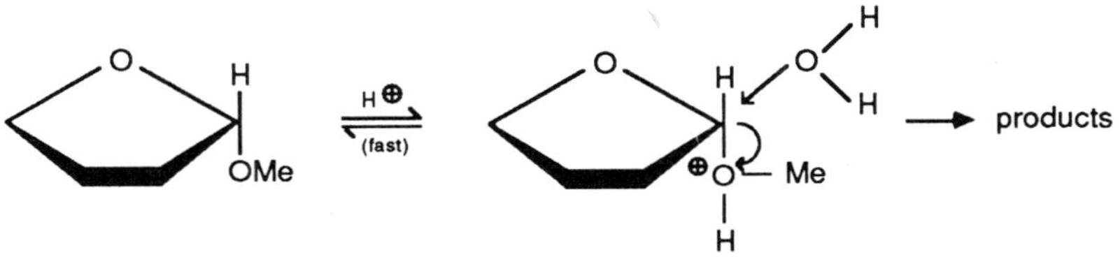

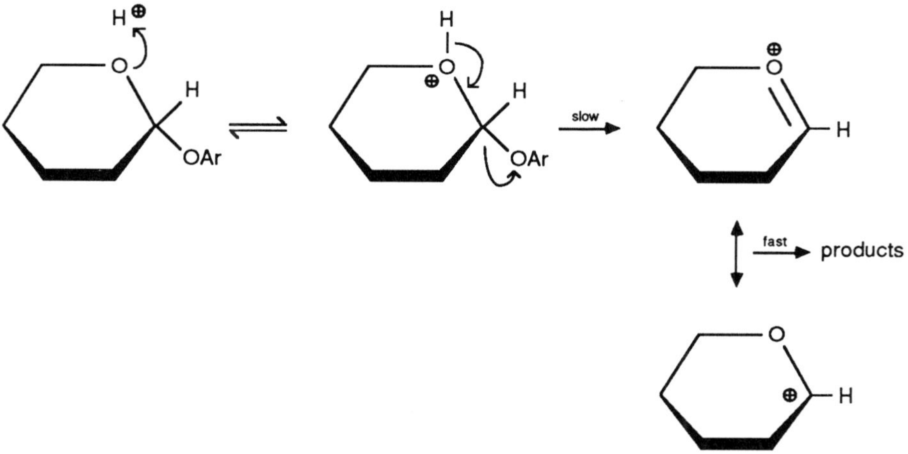

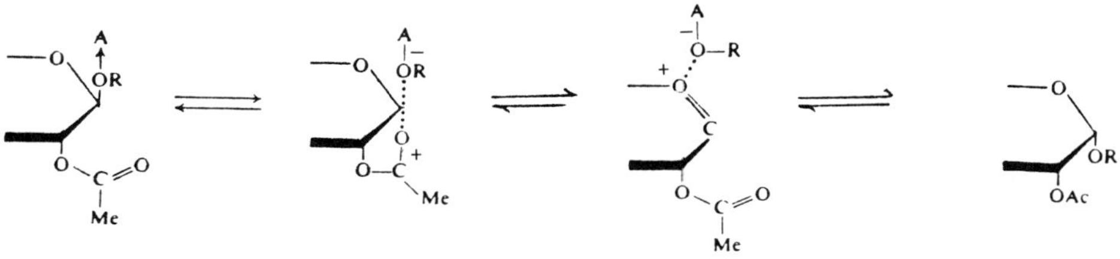

c. *Transglycosidation (OR→ OR′).* Transglycosidation is the exchange of the aglycon group (OR) of a glycoside with the OR′ group of another alcohol. The reaction is quite similar to the anomerization previously discussed, as in both cases the anomer having a bulky group in the axial position preponderates. The reaction may be conducted with such protic acids as HCl or such aprotic acids as $ZnCl_2$. Because oligosaccharides and polysaccharides are polymeric acetals linked by glycosidic bonds, methanolysis (transglycosidation with methanol) has often been used to prepare the methyl glycosides of the monomers. Thus, methyl α-D-mannopyranoside is conveniently prepared by treating the polysaccharide D-mannan, found in the ivory nut, with methanolic HCl.

d. *Glycosyl Halide Formation (OR → X).* Like ester groups situated in the anomeric position, the acetal groups of glycosides can readily be replaced by halogens. The reaction is usually performed by blocking a methyl glycoside with ester or other groups and then treating the product with anhydrous HCl or HBr. This method is useful when furanosyl halides are needed, because the necessary methyl glycofuranosides are accessible by the Fischer method of glycosidation. On the other hand, when glycopyranosyl halides are required, the esters are used as starting materials.

Glycosyl halides are used in the preparation of complex glycosides and 1-thioglycosides. For complex glycosides the Koenigs–Knorr method is used; for phenyl and thioglycosides the sodium salt of the phenol or thiol is caused to react directly with the halide without any catalyst.

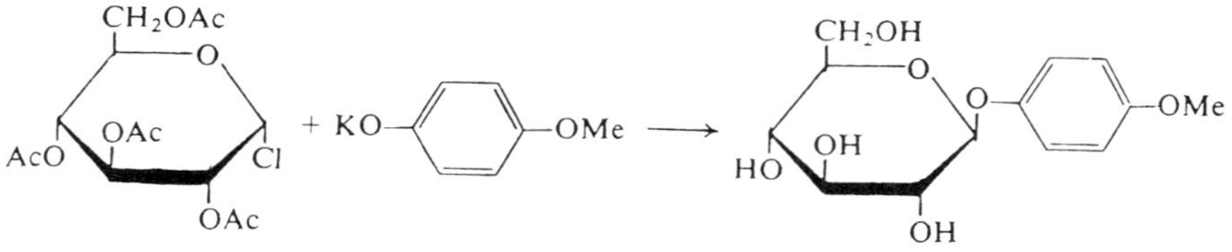

Glycosyl halides readily react with alcohols in the presence of Lewis acids. For example, tetra-*O*-acetyl-α-D-glucopyranosyl bromide reacts with methanol in the presence of $SnCl_4$ to yield the β-D-glucopyranoside [see S. Hanessian and J. Banoub, *Carbohydr. Res.* **44,** C14 (1975); **59,** 261 (1977)]. Alternatively $MeOSnBu_3$ may be reacted directly with the glycosyl halide to yield the 1,2-trans glycosides [see T. Ogawa and M. Matsui, *Carbohydr. Res.* **51,** C13 (1976); **56,** C1 (1977); **62,** C1 (1976)]. Glycosidation using $SnCl_4$ has also been carried out with esters [see El Khadem and D. Matsuura, *Carbohydr. Res.* **101,** C1 (1981)].

C. Nucleophilic Substitutions on Nonanomeric Carbon Atoms

For steric reasons, the addition of a nucleophile to an aldehydic group is more facile than its addition to a more crowded keto group. Accordingly, aldoses are more susceptible to nucleophiles than are ketoses, and the terminal carbonyl group of a glycosulose, i.e., the aldehydic group, is more reactive than the keto group. The reactivities of the two keto groups of a diulose are quite similar, and cyclization usually involves the keto group that is capable of forming a five- or six-membered hemiacetal ring with the oxygen atom of a free hydroxyl group. In general, whenever a nucleophilic attack occurs on the carbonyl group of a sugar, two epimeric addition compounds are isolated, and the more stable of the two isomers is the one that preponderates. Use is made of this stereoselectivity in the preparation of sugar derivatives of a desired configuration.

Unlike the addition reactions just discussed—where the susceptibility of carbonyl groups to nucleophiles changes slightly—when the carbonyl groups are situated terminally, or at one of the other positions, nucleophilic substitution reactions at the anomeric center are orders of magnitude more facile than substitution at the other carbon atoms of a sugar. This is attributed to the ability of the anomeric carbon atom to form resonance-stabilized carbonium ions, which render nucleophilic substitution reactions at the anomeric center much easier than at the other carbon atoms of the sugar molecule. The latter are relatively inert toward nucleophiles and react only if good leaving groups are linked to them. Examples of good leaving groups that may be used to increase the sensitivity of carbon atoms toward nucleophiles are the halogens, particularly bromine and iodine, as well as the tosyl, mesyl, brosyl, and triflate groups [see L. D. Hall and D. C. Miller, *Carbohydr. Res.* **40,** C1 (1975); **47,** 299 (1976)]. It should be noted that, when these substituents are replaced by nucleophiles, inversion of configuration usually takes place in the absence of a participating group, as seen in the following example.

Me_2C OCH_2 OCH O OH O $O—CMe_2$ $\xrightarrow{TsCl}$ Me_2C OCH_2 OCH O OTs O $O—CMe_2$ $\xrightarrow{NH_2NH_2}$ Me_2C OCH_2 OCH O H_2NNH O $O—CMe_2$ $\xrightarrow{H_2}$ Me_2C OCH_2 OCH O H_2N O $O—CMe_2$

(also from the tosylate with NH_2NH_2:) $H_2C—C(H)$ O O CMe_2 O O $O—CMe_2$

Because of the basic nature of hydrazine, the reaction is accompanied by some elimination. With ammonia, which is a stronger base and weaker nucleophile, more elimination occurs and very little substitution product results.

D. Reactions of the Hydroxyl Groups

There are two types of hydroxyl groups in the acyclic forms of monosaccharides: a primary hydroxyl group, which is, by definition, always terminal, and a number of secondary hydroxyl groups. In the cyclic forms of sugars there exists a third type, namely the glycosidic hydroxyl group attached to the anomeric carbon atom, which forms part of the hemiacetal group of aldoses or the hemiacetal group of ketoses. All three types of hydroxyl groups are strong nucleophiles that can add to the carbonyl group of acid anhydrides to form esters, or can replace the leaving groups of alkylating agents to form ethers. They can also induce substitution reactions at the anomeric center of the same or a different sugar to form anhydro derivatives or disaccharides, respectively. Finally, they can add in pairs to the carbonyl group of aldehydes and ketones to give cyclic acetals.

The hemiacetal hydroxyl groups of cyclic sugars are the most reactive of the three types of hydroxyl groups. They are followed in nucleophilicity by the terminal primary hydroxyl groups. Because acyclic sugars do not possess hemiacetal hydroxyl groups, their primary hydroxyl groups are the most reactive ones in the molecule. The least reactive hydroxyl groups in both cyclic and acyclic sugars are the secondary ones. These may differ in reactivity, depending on whether they are axially or equatorially oriented (eliminations are favored by an antiperiplanar orientation) and whether or not the substituents are situated in a crowded environment. Thus, it is often possible to block a primary hydroxyl group and leave the secondary hydroxyl groups free by making use of the fact that secondary hydroxyl groups, being in a more crowded environment, may not react to any appreciable extent if mild reaction conditions and insufficient reagents are used. Stereochemical considerations also play important, and sometimes decisive, roles in determining the course of competing reactions involving more than one hydroxyl group. For example, when methyl β-D-ribofuranoside, which possesses one primary and two secondary hydroxyl groups, reacts with acetone, the 2,3-*O*-isopropylidene derivative is preferentially formed by attack of the two cis-oriented secondary hydroxyl groups, despite the fact that a primary hydroxyl group (which is trans-oriented) is available. Had the latter group reacted, it would have produced a highly strained six-membered 3,5-*O*-isopropylidene ring.

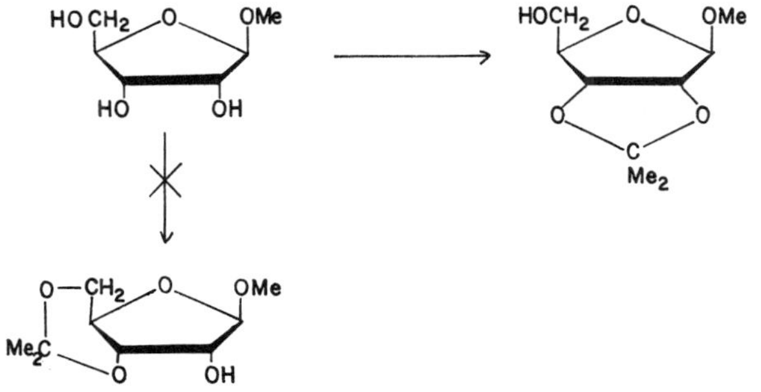

Because the OH-2 is the most acidic of the secondary hydroxyl groups attached to a saccharide ring (proximity to the anomeric carbon), it can be selectively esterified or etherified.

The foregoing generalizations are useful when planning multistage syntheses of complex sugar molecules, especially when it is advantageous to block certain hydroxyl groups selectively and leave the others free for subsequent reaction. They are also useful in predicting which group will preferentially react when a limited amount of reagent is available.

In the following subsections, four types of hydroxyl group derivatives will be discussed: (1) esters, (2) ethers, (3) anhydro sugars and disaccharides, and (4) cyclic acetals.

1. Formation of Esters

Esters are generally used to block hydroxyl groups, i.e., to deactivate their oxygen atoms and, by so doing, prevent them from attacking nucleophile acceptors. The esters most commonly used for this purpose are the acetates and benzoates. Occasionally, substituents are introduced in the phenyl ring of the latter esters to increase their crystallizing properties. The *O-p*-nitrobenzoyl and the *O-p*-toluoyl derivatives have been found to be useful in this respect.

Peracetylation (full acetylation) may be achieved at room temperature by treating the saccharide in pyridine with acetic anhydride, or at a higher temperature by heating the saccharide in a mixture of acetic acid and acetic anhydride. In both cases, the thermodynamically favored pyranose derivative is obtained. If the furanose derivative is desired, the methyl furanoside is acetylated and the product is subjected to acetolysis (hydrolysis and acetylation) to replace the OMe group by OAc. The last reaction is conducted at low temperature with a mixture of acetic acid, acetic anhydride, and a few drops of sulfuric acid.

To prepare benzoates, *p*-substituted benzoates, and sulfonates, the necessary acid chloride is allowed to react in pyridine with the saccharides or saccharide derivatives.

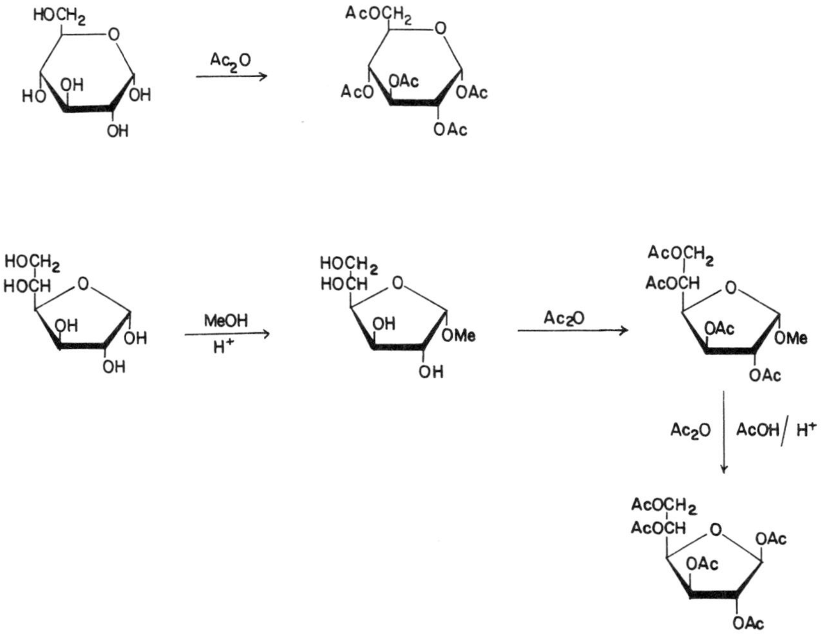

Formation of peracetylated pyranoses and furanoses

In general, ester groups are more stable in acidic than in basic media, and acetates are more readily hydrolyzed than benzoates. To carry out a deacetylation, a solution of sodium methoxide is added in catalytic amount to the sugar acetate at 0°C. Most esters are also saponifiable with NaOH in acetone at low temperature. Acetyl, benzoyl, and *p*-nitrobenzoyl groups attached to the oxygen atom of the anomeric centers of cyclic sugars can be displaced by nucleophiles, especially in the presence of Lewis acid catalysts. This reaction is useful in the preparation of glycosides and nucleosides. Ester groups attached to the other carbon atoms of a saccharide, i.e., linked to nonanomeric carbon atoms, are considerably more inert toward nucleophiles and will not undergo substitution reactions under mild reaction conditions. To achieve a nucleophilic substitution of esters attached to nonanomeric centers, the esters must themselves be good leaving groups, for example, tosylates, mesylates, and triflates (see equation, p. 137).

2. Formation of Ethers

In true saccharide ethers the hydroxyl groups attached to the nonanomeric carbon atoms are replaced by alkoxyl groups. These derivatives should be distinguished from the structurally similar alkyl glycosides, discussed earlier, which possess acetal-type OR groups. Methyl ethers have been used extensively in the elucidation of structure of saccharides. They were first used by Haworth to determine the ring size of monosaccharides and the ring size and position of linkage of oligosaccharides. More recently, they have been used to volatalize monosaccharides before subjecting them to gas chromatographic analysis. Because silyl derivatives are easier to prepare than methyl ethers, they have now replaced the methyl ethers in the gas chromatographic analysis of monosaccharides. On the other hand, because methyl ethers are stable toward acid- and base-catalyzed hydrolysis, they continue to be used as a means of labeling free hydroxyl groups in saccharides, a procedure frequently used in the structure elucidation of oligosaccharides and polysaccharides. For this reason the problem of permethylating saccharides (etherifying all of their hydroxyl groups) continues to attract the attention of carbohydrate chemists. The original methods of Purdie (MeI and AgOH) and Haworth (Me_2SO_4 in alkali) have been much improved by the use of such aprotic solvents as $HCONMe_2$ (DMF) or Me_2SO. Today, the most widely used methylating procedures are the Hakomori methylation, which uses sodium hydride and Me_2SO to generate the base $MeSOCH_2^-$, and methylation with sodium hydride–*N,N* dimethylformamide and alkyl halides. Addition of tetrabutylammonium iodide in tetrahydrofuran increases the reactivity of hindered R-OH groups by forming a reactive species.

Benzyl ethers offer unique advantages in syntheses requiring selective blocking and deblocking of hydroxyl groups. They can be introduced under mild conditions by the action of benzyl chloride and NaH in DMF and can be removed in neutral media by catalytic hydrogenolysis, which does not affect esters or cyclic acetals. The former class of compounds is base-labile and the latter acid-labile.

R = $C_6H_5CH_2$

Other synthetically useful ethers are the triphenylmethyl ethers or, as they are often called, trityl derivatives (Ph_3C—O—). Their bulky phenyl groups render their formation from secondary and tertiary hydroxyl groups so difficult that they are normally obtained from primary hydroxyl groups only. These ethers are, therefore, used to block the primary hydroxyl groups selectively, to leave the secondary hydroxyl groups free for subsequent reactions. Removal of the trityl ethers is very facile and can be achieved by mild acid hydrolysis. Organic acids that do not affect esters or cyclic acetals, such as acetic acid, selectively deblock the oxygen atom bearing the trityl group and regenerate the primary hydroxyl group with liberation of triphenylmethanol. Alternatively, the trityl ether group may be removed by catalytic hydrogenolysis, yielding triphenylmethane and the free primary hydroxyl group.

The ease of hydrolysis of trityl ethers may sometimes be responsible for the occurrence of undesirable hydrolysis during a subsequent reaction, particularly when vigorous conditions are used.

HOCH2 R, HO —TrCl→ TrOCH2 R, HO —MsCl→ TrOCH2 R, MsO —NaI→ —H_2/Pd→ HOCH2 R

Tr = triphenylmethyl

It is also possible to prepare trityl ethers of secondary hydroxyl groups when these are required [see Y. V. Wozney and N. K. Kochetkov, *Carbohydr. Res.* **54,** 301 (1977)].

3. Anhydrides and Disaccharides

An intra- or intermolecular nucleophilic attack initiated by the oxygen atom of a suitably placed hydroxyl group on the anomeric carbon atom of the same or a different sugar will afford an anhydride or a disaccharide, respectively. The reaction is best performed by introducing a good leaving group, usually a halogen atom, on the anomeric carbon atom and blocking all of the hydroxyl groups except the one to be involved in the subsequent reaction. In this way, the formation of undesired products can be avoided (see p. 143).

4. Cyclic Acetals

If oriented properly, any two adjacent (but not necessarily contiguous) hydroxyl groups will react with an appropriate aldehyde or ketone to yield an unstrained five-membered or six-membered cyclic acetal. The most

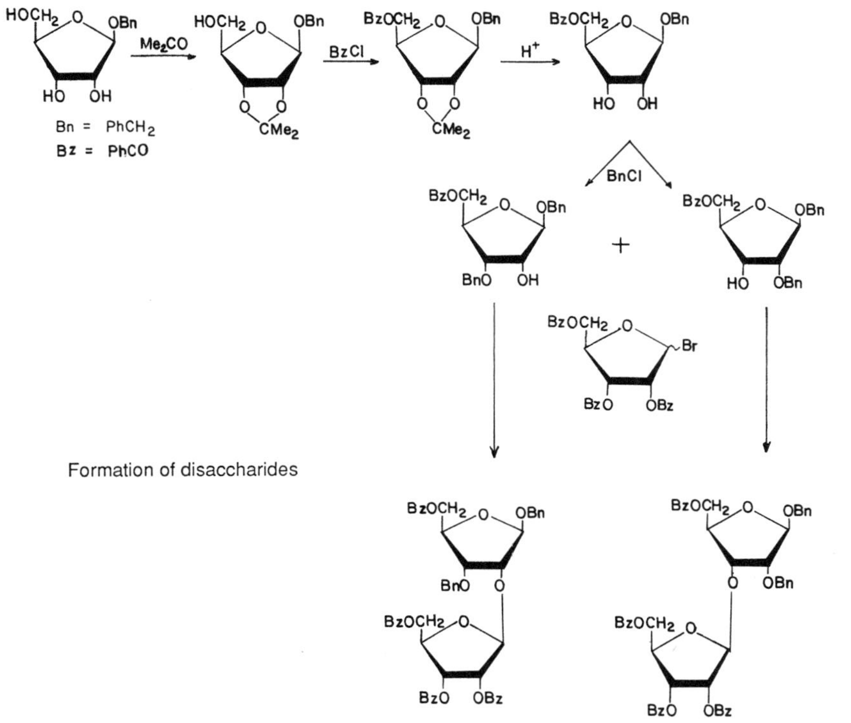

Formation of disaccharides

commonly used carbonyl compounds are benzaldehyde, which yields five- and six-membered benzylidene acetals, and acetone, which yields almost exclusively five-membered rings (isopropylidene acetals). The latter derivatives have the advantage of existing in one isomeric form, unlike benzylidene derivatives, which exist in two diastereomeric forms. On reacting with chiral glycols, all aldehydes, except formaldehyde, and all mixed ketones yield chiral diastereomeric acetals.

The formation of cyclic acetals is catalyzed by acids and proceeds by successive nucleophilic attack of one hydroxyl group on the protonated carbonyl derivative to form the hemiacetal. The latter, in turn, becomes protonated and is attacked by the second hydroxyl group.

Isopropylidene acetals may be formed by treatment of saccharide derivatives with acetone or its dimethyl acetal (2,2-dimethoxypropane) in the presence of an acid catalyst such as HCl or H_2SO_4 and a dehydrating agent. The reaction is often conducted at room temperature and may be monitored by ^{1}H-NMR spectroscopy. The signals of the two methyl groups are usually well resolved because of their chiral environment. [For a study of acetonation of carbohydrates with 2-alkoxypropenes, see J. Gelas and D. Horton, *Heterocycles* **16,** 1587 (1981).]

A useful starting material in many syntheses is 1,2 : 5,6-di-*O*-isopropylidene-α-D-glucofuranose, obtained by treating D-glucose with acetone. The derivative provides an easy means of access to furanoses having a free hydroxyl group on C-3 (see p. 137).

Benzylidene groups are also useful for selective blocking of saccharide derivatives. They usually involve the primary hydroxyl group of a pyranoside and the 4-hydroxyl group if it is suitably positioned. Benzylidene groups are also useful for introducing a halogen atom instead of the primary hydroxyl group by the Hanessian method using *N*-bromosuccinimide (NBS), as shown in the following sequence of reactions carried out by Horton.

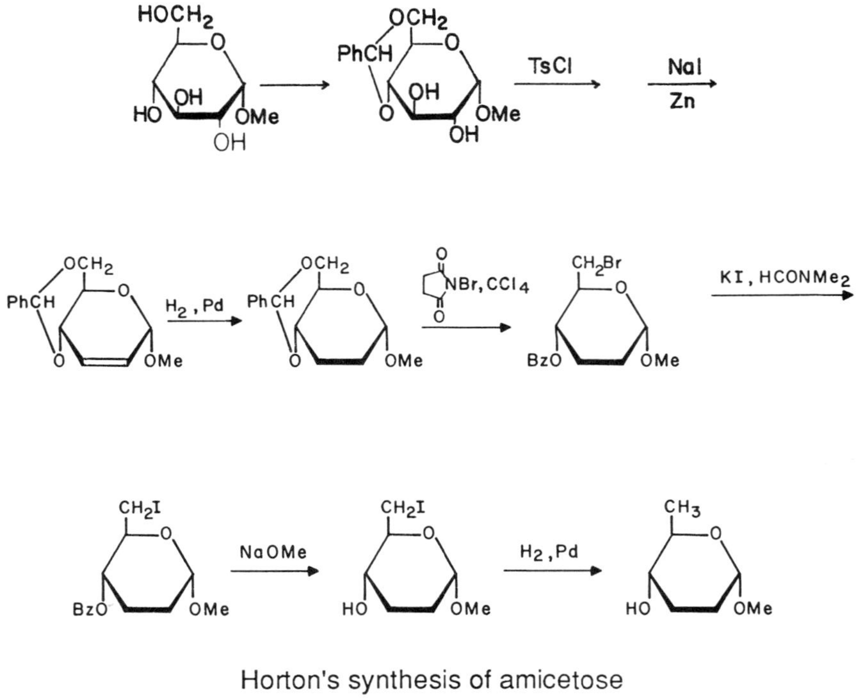

Horton's synthesis of amicetose

E. Oxidation and Reduction

1. Oxidation

Oxidation of a sugar involves breaking of carbon–hydrogen, carbon–carbon, or oxygen–hydrogen bonds and transfer of electrons to the oxidant. Two types of oxidation are possible. The first involves transfer of two electrons from one atom to the oxidant and is referred to as heterolytic oxidation. The other, known as homolytic oxidation, occurs in two steps, each involving the transfer of one electron. The two types of oxidation will be discussed separately, but there will be some overlap as some oxidations can proceed by both mechanisms, whereas others have not been sufficiently studied to permit a decision as to the mechanism of their oxidation.

a. *Heterolytic Oxidations.* Heterolytic oxidations, i.e., those involving two-electron transfers, may be exemplified by the oxidation of a secondary hydroxyl group to a keto group. This involves the elimination of two hydrogen atoms and the formation of a double bond between the carbon atom and the oxygen atom of the alcohol function. The first step takes place by breakage of a C–H bond and elimination of a hydride ion. Because the hydride anion is a very poor leaving group, this bond breakage is generally the slowest step of the oxidation (the rate-determining step). What makes the reaction possible, despite this, is the fact that the elimination is aided by the oxidant, which captures the pair of electrons and converts the hydride ion into a proton. Another proton is concurrently liberated by dissociation of the O–H bond of the alcohol, and both protons react with a base (usually a hydroxyl group to form water). Such an oxidation usually occurs by a concerted or E2 type of mechanism, in which the oxidant is not bonded to the sugar, except possibly in the transition state.

$$\text{(Oxidant)} + \text{H—}\overset{|}{\underset{|}{\text{C}}}\text{—O—H} \rightarrow \text{(oxidant)}^{2-} + \overset{|}{\underset{|}{\text{C}}}\text{=O} + 2\ \text{H}^{2+}$$

Some oxidants, for example, chromic acid, have been shown to form intermediate esters, which subsequently decompose by bimolecular eliminations. The difference between this and the previous oxidation is that the leaving group is the reduced form of the oxidant, and the capture of electrons by the oxidant is the driving force of the reaction. Furthermore, the breaking of the C–H bond, which occurs simultaneously, is the rate-determining step.

$$\text{(Oxidant)—O—}\overset{|}{\underset{|}{\text{C}}}\text{—H} + \text{:(base)} \rightarrow \text{(oxidant)}^{-} + \overset{|}{\underset{|}{\text{C}}}\text{=O} + \text{(base)H}$$

Glycols are more acidic than monohydric alcohols, and the C-1 group on an aldopyranose is even more acidic than either, because of the inductive effect of the ring oxygen atom. For this reason, sugars and glycosides are more readily oxidized than ordinary alcohols. Oxidation of free sugars at higher pH is often accompanied by competing processes of epimerization and degradation. In general, the β anomers of D sugars and their glycosides are more rapidly oxidized than the α anomers (for an explanation, see the discussion of the anomeric effect). Similarly, the 2-hydroxyl group of methyl β-D-glucopyranoside is more acidic than the corresponding group in the α-D anomer.

(i) HALOGEN AND HYPOHALITE OXIDATIONS. Halogens (usually bromine) and hypohalites (particularly sodium hypoiodite) have been used to oxidize aldoses to aldonic acids and their lactones. The aldehydic group of acyclic sugars is converted into a carboxyl group, and the anomeric hydroxyl group is converted into a lactone. Although the terminal (primary) hydroxyl groups of sugars can also undergo oxidation by these reagents, the reaction occurs at a somewhat lower rate, so the oxidation may be stopped at the monocarboxylic acid (or lactone) stage. By using moderately vigorous conditions, oxidations of the primary hydroxyl groups have been used to convert glycosides into glycosiduronic acids. The oxidation of secondary hydroxyl groups to keto groups requires drastic conditions, and these may be accompanied by cleavage of carbon–carbon bonds.

It should be noted that changes in temperature, acidity, and concentration can lead to changes in the nature of the oxidizing species. For example, in acidic solution, the equilibrium between free halogen and hypohalous acid lies far to the left and the concentration of hypohalous acid is very low, but when sufficient alkali is added to the system, the concentration of hypohalite ion increases dramatically. For example, at pH 1, 82% of the total chlorine present in solution exists as free chlorine and 18% as hypochlorous acid. At pH 4, only 0.4% of the chlorine is free and 99.6% exists as hypochlorous acid, whereas at pH 8, 21% exists as hypochlorous acid and 79% as hypochlorite.

$$X_2 + H_2O \rightleftarrows HOX + HX$$

$$X_2 + 2\ OH^- \rightleftarrows OX^- + X^- + H_2O$$

(ii) OXIDATION IN ACIDIC SOLUTIONS. In acidic solutions, the active oxidant is normally the free halogen molecule, and the hypohalous acid plays a minor role. The order of effectiveness of halogens as oxidants is $Br_2 > Cl_2 > I_2$, with the last being quite ineffective. This order corresponds to the rate of reaction of the free halogen with water and to its solubility in water, which explains the ineffectiveness of free iodine as an oxidant.

It should be noted that, unless a buffer or neutralizing compound is present during the oxidation of a free sugar, the solution becomes strongly acidic as a result of the formation of hydrohalic acid.

$$\text{R-CHO} + Br_2 + H_2O \rightarrow \text{R-COOH} + 2HBr$$

$$\text{R-CHO} + HOBr \rightarrow \text{R-COOH} + HBr$$

The accumulation of hydrogen bromide during oxidations by bromine profoundly inhibits the rate of further oxidation. This effect is not due only to the simple increase in acidity; although other strong acids also inhibit the rate, the effect is largest for hydrogen bromide and chloride. To minimize this inhibiting influence, the reaction may be conducted in the presence of a solid buffer such as barium carbonate. In general, the presence of a buffer increases the yield of aldonic acid and precludes the hydrolysis of disaccharides. For D-gluconic acid, yields of 96% have been reported.

When the oxidation period is extended for unbuffered solutions, keto acids may be formed in small yields. Thus, hexoses give hex-5-ulosonic acids. Under more drastic conditions carbon–carbon bonds are cleaved, yielding chain-shortened acids.

Commercial production of aldonic acids has been achieved by electrolysis of dilute solutions of sugars in the presence of a bromide and a solid buffer such as calcium carbonate. Presumably, free bromine forms at the anode and then oxidizes the aldose to the aldonic acid and is itself reduced to bromide. If the electrolytic method is not well controlled, aldaric acids and glyc-2- and -5-ulosonic acids may be produced.

Ketoses are generally resistant to bromide oxidation, and the latter may be useful in removing aldose contaminants from ketoses. However, when the period of oxidation is extended, D-fructose yields D-*lyxo*-5-hexulosonic acid.

The mechanism of the oxidation of aldoses with chlorine and bromine in acid media was studied by Isbell, who found that the active oxidants are the halogens and not the hypohalous acids. For example, molecular chlorine was found to be the active oxidant in the oxidation of D-glucose by buffered chlorine–water at pH 3. Interestingly, the cyclic forms of an aldose, but not the free aldehyde, are oxidized directly under these conditions. Pyranoses give 1,5-lactones, and furanoses 1,4-lactones, directly and in high yields. As might be expected, the equilibrium solutions are oxidized at rates intermediate between those for the individual anomers. It is also interesting that oxidation of β-D-glucopyranose is 250 times as fast as that of the α-D anomer (see anomeric effect, p. 53). Overend suggested that the rates observed for the oxidation of α-D anomers are actually the rates of mutarotation. An explanation for the higher rate of oxidation of β-D-glucopyranose is the equatorial orientation of its 1-hydroxyl group, which renders it much more accessible to the oxidant. It is also facilitated by overlap of the lowest unoccupied molecular orbital (LUMO) of the axial hydrogen with the highest occupied molecular orbital (HOMO) of the ring oxygen atom, which would ease the removal of the hydrogen as H^-. The conjugate acid formed loses a proton, and the free hypobromous ester undergoes an E2 elimination to form D-glucono-1,5-lactone. A similar elimination from α-D-glucopyranosyl hypobromite would occur from a conformation destabilized by crowding between the bromine molecule and the protons axially attached to C-3 and C-5.

D-Glucono-1,5-lactone

α- and β-Glucopyranose

An alternative explanation, proposed by Isbell, is based on the difference in free energy between the ground and transition states of the two anomeric anions derived from D-glucopyranose. The β-D anomer suffers less destabilization by nonbonded interactions than the α-D anomer. It is interesting that higher rates of oxidation are also observed for methyl β-D-glucopyranoside than for the α-D anomer, and in this case no hypohalite ester can be formed at the anomeric carbon atom.

In alkaline solution, halogens are converted into hypohalous acids and hypohalite ions. The latter are stronger oxidants than free halogens, which explains why, for example, iodine does not oxidize sugars, whereas hypoiodite is a powerful oxidant. Hypobromite and hypochlorite oxidize primary and secondary alcoholic groups and can cause cleavage of carbon–carbon bonds. Hypohalite oxidation is complicated by the tendency of hypohalites to disproportionate to halate ions, which readily oxidize aldonic acids and their lactones to the glyc-2-ulosonic acids. For example, D-glucono-1,4-lactone in the presence of vanadium pentaoxide is oxidized by chloric acid to methyl D-*arabino*-2-hexulosonate.

CH$_2$OH, HOCH, O, OH, H, =O, H, H, OH $\xrightarrow[V_2O_5,\ MeOH]{HClO_3}$ CO$_2$Me, C=O, HOCH, HCOH, HCOH, CH$_2$OH

(iii) NITRIC ACID AND NITROGEN DIOXIDE. Nitric acid, a strong acid, is a potent oxidant that converts primary alcoholic and aldehydic groups into carboxylic acid groups. Kiliani, a contemporary of Emil Fischer, was the first to study the oxidation of carbohydrates with nitric acid. He found that aldoses are oxidized to dicarboxylic acids; for example, D-galactose yields galactaric (mucic) acid. Under similar conditions, alditols are oxidized to aldonic acids, and aldonic acids to a mixture of 2-glyculosonic acids, aldaric acids, and alduronic acids. It will be recalled that oxidation of methylated sugars with nitric acid was used by Haworth to determine the position of unsubstituted hydroxyl groups in mono- and disaccharides. In these oxidations, cleavage of the carbon–carbon bonds leads to the formation of polyhydroxyglutaric acids as well as of tetraric and oxalic acids.

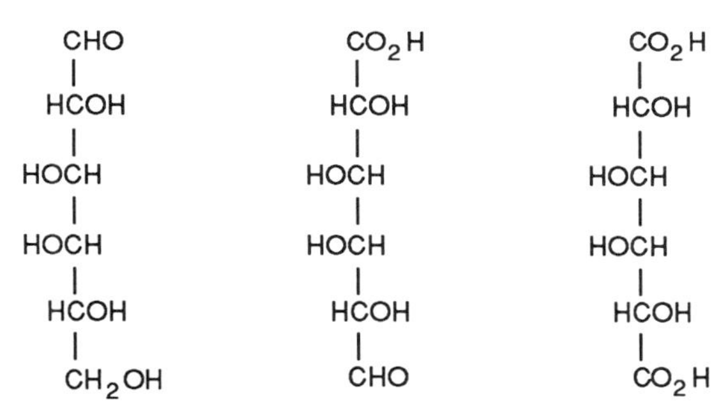

The specificity of the reaction may be enhanced by the use of nitrogen dioxide instead of nitric acid. Bubbling this gas into a solution of a methyl glycoside converts it into the methyl aldosiduronic acid in good yield.

(iv) OXIDATION WITH RUTHENIUM TETRAOXIDE. Ruthenium tetraoxide is a powerful oxidant that combines explosively with ether or benzene; for this reason it is generally used in dilute solutions in carbon tetrachloride. The oxidant is prepared in the stoichiometric amount needed before the reaction or generated *in situ* by reaction of a catalytic amount of ruthenium dioxide with periodate or hypochlorite.

Ruthenium tetraoxide can selectively oxidize primary hydroxyl groups to carboxylic acids. For example, 1,2-*O*-isopropylidene-α-D-xylofuranose is converted into the corresponding alduronic acid derivative. Use is made of the fact that isopropylidene groups, as well as acetates and benzoates, are inert to this oxidant. Under more drastic conditions, unprotected secondary hydroxyl groups will yield keto groups (see below).

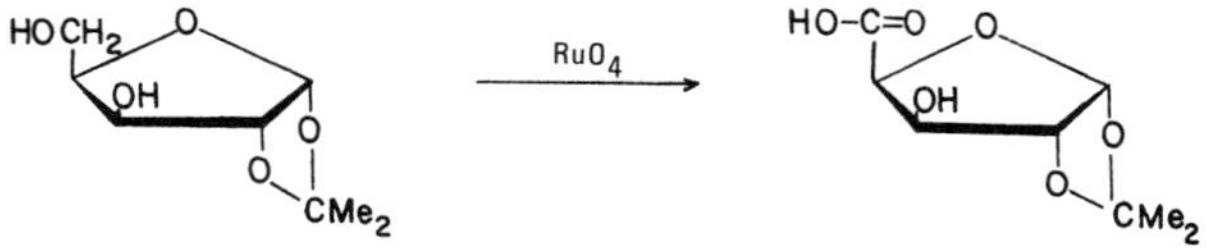

HOCH2 — OH — O — CMe2 —RuO_4→ HO–C=O — OH — O — CMe2

(v) OXIDATION WITH DIMETHYL SULFOXIDE (DMSO). Apart from the Pfitzner–Moffatt oxidation, which uses DMSO and dicyclohexylcarbodiimide (DCC), other activating agents such as acetic anhydride, phosphorus pentaoxide, boron trifluoride and the sulfur trioxide–pyridine complexes have been used with DMSO to oxidize primary hydroxyl groups to aldehydes and secondary hydroxyl groups to ketones. [For a good review see G. H. Jones and J. G. Moffatt, *Methods Carbohydr. Chem.* **6,** 315 (1972).]

Thus, oxidation of 1,2 : 3,4-di-*O*-isopropylidene-α-D-galactopyranose yields a dialdose, and oxidation of 1,2 : 5,6-di-*O*-isopropylidene-α-D-glucofuranose yields the 3-ulose.

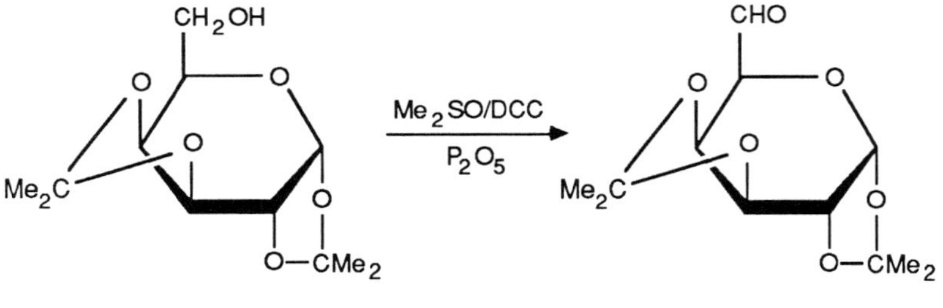

CH_2OH, Me_2C, O–CMe_2 —Me_2SO/DCC, P_2O_5→ CHO, Me_2C, O–CMe_2

Me_2C OCH_2 / OCH, OH, O–CMe_2 —Me_2SO/P_2O_5→ Me_2C OCH_2 / OCH, O, O–CMe_2

Because primary hydroxyl groups react faster than secondary hydroxyl groups, it has been possible to oxidize unprotected derivatives. However, the method has been plagued by problems, such as the side reactions that often occur during the oxidation. For example, acetic anhydride and dimethyl sulfoxide may give (methylthio) methyl ethers instead of oxidizing the sugar derivative. Acetic acid is sometimes lost by elimination from partially acetylated derivatives, producing a double bond conjugated with the newly formed carbonyl group. Another problem is difficulty in removing the high-boiling dimethyl sulfoxide from the reaction mixture, which necessitates lyophilization unless the products have low solubility in water.

b. *Homolytic Oxidations.* Most homolytic oxidations occur in two successive steps, each of which involves a one-electron transfer. The more difficult (and slower) step in the homolytic oxidation of a sugar is the first, in which a hydrogen atom is abstracted from a C–H group, followed by transfer of an electron from this hydrogen atom to the oxidant to yield a proton. In this process, the sugar that lost a hydrogen atom is converted into a radical, which is stabilized by resonance (the unshared pair of electrons on the adjacent oxygen atom can migrate to the carbon atom from which hydrogen was abstracted). This explains why the initial abstraction of hydrogen usually takes place by removal of a hydrogen atom attached to carbon (H—|—C—O—H) rather than the one attached to oxygen (H—C—O—|—H). It is also in agreement with the fact that compounds having equatorial hydrogen atoms linked to carbon atoms (i.e, carbon-linked hydrogen atoms, which are more accessible to the oxidant than the axially oriented hydroxyl groups attached to the same carbon atom) undergo more facile catalytic oxidation than axially oriented C–H groups (which are linked to equatorial hydroxyl groups). The final step of the oxidation is a fast reaction involving homolysis of the O–H bond of the sugar radical to give a carbonyl group plus a second hydrogen radical, which, like the first, is converted by the oxidant into a proton.

$$\mathrm{Ox} + \mathrm{H{-}\overset{|}{\underset{|}{C}}{-}O{-}H} \longrightarrow \mathrm{Ox^-} + \mathrm{H^+} + \mathrm{\cdot\overset{|}{\underset{|}{C}}{-}O{-}H}$$

$$\mathrm{\cdot\overset{|}{\underset{|}{C}}{-}O{-}H} \longleftrightarrow \mathrm{{}^-\overset{|}{\underset{|}{C}}{-}O^+{-}H}$$

$$\mathrm{Ox} + \mathrm{\cdot\overset{|}{\underset{|}{C}}{-}O{-}H} \longrightarrow \mathrm{\overset{|}{\underset{|}{C}}{=}O}$$

(i) OXIDATION WITH MOLECULAR OXYGEN. It is now established from electron spin resonance (ESR) studies that oxidations with oxygen and with hydrogen peroxide proceed via a free-radical mechanism. This is why both oxidations lead to somewhat similar products. Like all free-radical reactions, they are started by initiation reactions, brought to completion by chain propagation reactions, and then concluded by termination reactions. The last are of little concern, because there is always an excess of the reagents.

Initiation reactions: In its ground state, the oxygen molecule exists as a stable diradical whose reactivity is greatly enhanced by the presence of such catalysts as palladium or platinum or of OH^- groups. Free radicals (R·) are usually generated by initiation reactions involving abstraction of hydrogen from an activated molecule (R*H) to form activated radicals ($R^{\dot{*}}$), which are then converted by oxygen into peroxidate radicals (R*-O-O·). When one of these radicals encounters a substrate molecule (RH), a hydrogen atom is abstracted, the molecule is converted into a radical (R·), and the peroxidate radical is converted into a hydroperoxide (R*-O-OH). Alternatively, the substrate molecule (RH) reacts with a hydroxyl group to yield a carbanion, which then reacts with another oxygen molecule to form the same radical (R·).

$$R^*H + O_2 = R^{\dot{*}} + HOO\cdot$$

$$R^{\dot{*}} + O_2 = R^*\text{-O-O}\cdot$$

$$RH + R^*\text{-O-O}\cdot = R\cdot + R^*\text{-O-OH}$$

$$RH + OH^- = R^- + H_2O$$

$$R^- + O_2 = R\cdot + \text{O-O}^-$$

Propagation reactions: These involve reactions of the radicals (R·) formed by one of the foregoing sequences of reactions with oxygen to form a peroxy radical (R-O-O·). On encountering the substrate (RH), this generates another radical (R·) and a hydroperoxidate (R-O-O-H), which disproportionates to an alkoxyl radical (R-O·) and a hydroxyl radical (H-O·). The former may then react with another substrate molecule to give an additional radical (R·) and R-OH.

$$R\cdot + O{-}O \longrightarrow R{-}O{-}O\cdot$$

$$+$$

$$RH \longrightarrow R\cdot + R\cdot{-}O{-}OH$$

$$\longrightarrow \cdot OH + R{-}O\cdot$$

$$+$$

$$RH \longrightarrow R{-}OH + R\cdot$$

In alkaline solution, oxygen degrades aldoses to aldonic acids having one carbon less than the starting sugar. Thus, when oxidized with oxygen (or air) in dilute KOH solution, D-glucose yields potassium D-arabinonate and potassium formate. Quantitative examination of the products formed shows that other acids such as erythronic, glycolic, glyceric, lactic, oxalic, and carbonic acid are formed in minor proportions. Isbell found that the reaction is initiated by the alkali, which converts the aldose or ketose into the 1,2-enediol. In the absence of oxygen, this intermediate undergoes further rearrangement by double-bond migrations, isomerization, degradation, and polymerization to produce a dark-colored reaction mixture. In the presence of oxygen, such coloration does not occur because the enediol quickly reacts with oxygen to form a diradical having two unpaired electrons whose spins are parallel (which prevents the two electrons from forming a bond). After a series of electron migrations, which may occur before or after spin reversal, a bond is formed between oxygen and carbon and formic acid is eliminated, giving the lower aldonic acid.

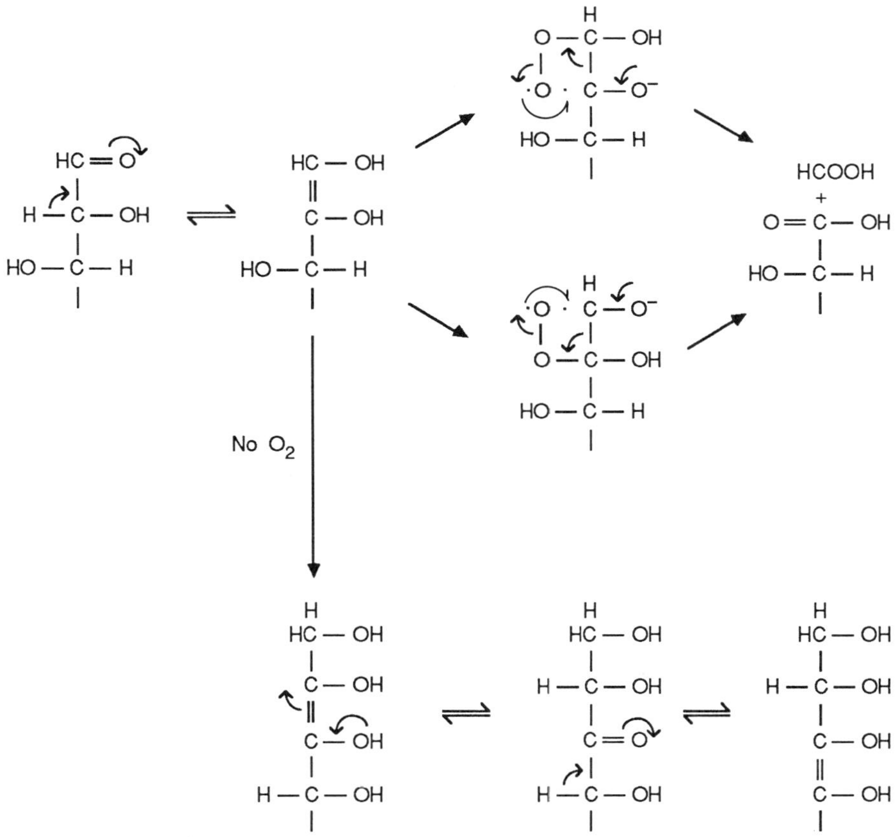

In neutral solutions, oxidation with air necessitates the presence of such catalysts as platinum, which causes dehydrogenation of the sugars. This is evident from the fact that isotopic oxygen is not incorporated in the acid formed and that the dehydrogenation is not reversible in the presence of $H_2{}^{18}O$. The radical process initiated by attack on a C–H bond (not an OH group, as in heterolytic oxidations) occurs on the catalyst surface and is strongly influenced by steric effects. For this reason, primary alcoholic groups are more readily oxidized than secondary ones. When the latter are oxidized, the attack is usually on a carbon atom bearing an axial hydroxyl group.

Oxidation with oxygen in the presence of platinum has been used to prepare a number of useful derivatives in good yields. For example, 1,2-*O*-isopropylidene-α-D-glucuronic acid was obtained from 1,2-*O*-isopropylidene-α-D-glucofuranose.

HOCH2 / HO-C-H / O / OH / O / O—CMe2 ⟶ HO—C=O / HO-C-H / O / OH / O / O—CMe2

(ii) OXIDATION WITH HYDROGEN PEROXIDE. Hydrogen peroxide is a very ineffective oxidant in acidic or neutral aqueous solutions, but in alkaline solutions or in the presence of catalysts it is quite effective. Such catalysts as ferrous or ferric salts enhance the oxidizing properties of hydrogen peroxide by promoting the formation of the free radicals ·OH and ·O-OH, respectively. The first (·OH) is by far the more effective oxidant. In the absence of free-radical catalysts, hydrogen peroxide has been used in alkaline media to oxidize aldoses to formic acid and the lower aldoses. As already discussed, under similar conditions of alkaline pH, oxygen yields the lower aldonic acid and formic acid. The difference is caused by the fact that the net effect in an oxygen oxidation is addition of O_2, whereas H_2O_2 ultimately dissociates to water, so that only one oxygen atom is added.

The hydrogen peroxide oxidation of aldoses in alkaline media usually involves hydroperoxide anions ($^-$O-OH), which may react with hydrogen peroxide to give hydroxyl radicals (·OH); these react with more peroxide anions to give hydroperoxide radicals (·O-OH). Addition of the peroxide anion to an aldose is illustrated. It may be noted that the formation of water can also be achieved with a hydroxyl radical, and in this case ·OH is eliminated to compensate for the consumption of hydroxyl radicals.

In general, ferric ions catalyze the formation of the hydroperoxy radical (·O-OH), and ferrous ions catalyze that of the hydroxyl radical (·OH).

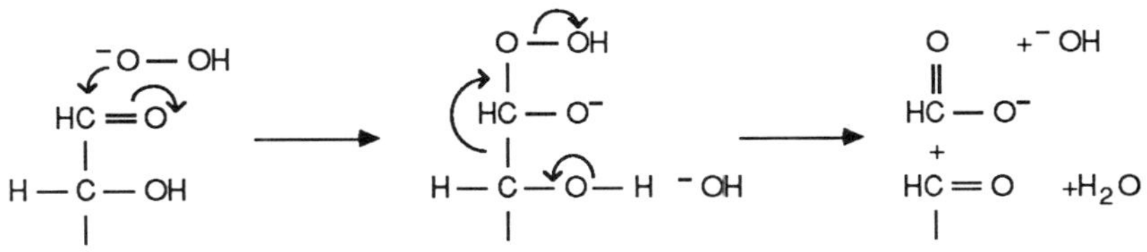

The former radical is the mild oxidant used in the Ruff degradation to convert aldonic acids into the next lower aldoses. As an oxidant, the hydroperoxy radical is not as effective as the hydroxyl radical, so the aldose formed from the higher aldonic acid can be isolated in reasonable yields. However, in the presence of an excess of hydrogen peroxide, the ferrous ions formed in solution catalyze the formation of hydroxyl radicals, which cause the reaction to proceed farther and thus lower the yield.

$$Fe^{3+} + HOOH + Fe^{2+} + H + \cdot OOH$$

$$Fe^{2+} + HOOH + Fe^{3+} + OH\cdot + OH$$

Under mild conditions:

Reaction with ferric complexes of aldonic acids:

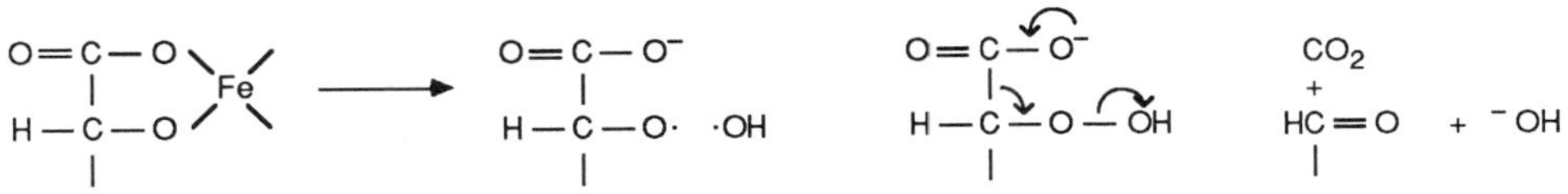

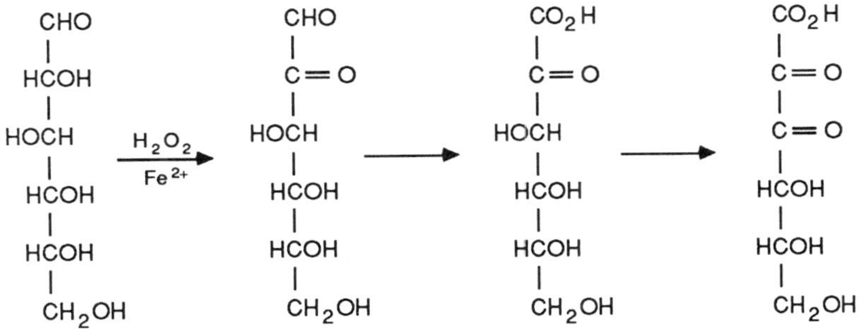

The Fenton reagent is a strong oxidant composed of hydrogen peroxide and ferrous sulfate. A free sugar treated with this reagent is degraded in a stepwise manner to formic acid. Under very mild conditions, the reaction may be stopped at the aldosulose stage. However, the yields are very low because the reaction proceeds further, yielding glyculosonic acids and diulosonic acids.

2. Reduction of Carbohydrates

Numerous reagents have been employed to reduce carbohydrates and their derivatives. These include inorganic hydrides such as those of aluminum and boron; reactive metals such as sodium in the presence of proton donors; and metal catalysts such as palladium, platinum, or nickel in the presence of molecular hydrogen.

a. *Lithium Aluminum Hydride.* Reductions of carbohydrate derivatives with lithium aluminum hydride are usually performed in nonaqueous solutions (such as benzene, ether, or tetrahydrofuran), because the reagent reacts violently with water. The boric acid formed during the borohydride reduction is usually removed as volatile methyl borate. Lithium aluminum hydride coverts carbonyl groups into hydroxyl groups and ether lactones into alditols. For example, 2,3,5,6-tetra-*O*-acetyl-D-glucono-1,4-lactone yields 2,3,5-tetra-*O*-acetyl-D-glucitol (the acetyl groups are reductively cleaved by the reagent).

$$>C{=}O + (AlH_4)^- \longrightarrow >C(H)(OAlH_3^-) \xrightarrow{H_2O} >C(H)(O^-) + Al(OH)_3$$

b. *Sodium Borohydride.* In contrast to lithium aluminum hydride, sodium borohydride and lithium borohydride are relatively stable in water and may be used to reduce lactones to sugars in acid media or to the alditol in basic media. Thus D-glucono-1,5-lactone yields D-glucose at pH 5 and D-glucitol at pH 9.

Sodium borohydride (or sodium borodeuteride) is also used to convert aldoses or oligosaccharides into alditol derivatives for examination by gas–liquid chromatography–mass spectrometry.

$$>C{=}O + [BH_4]^- \longrightarrow >C(H)(OBH_3^-) \longrightarrow >C(H)(O^-) + H_3BO_3$$

$$\text{D-glucono-1,5-lactone} \xrightarrow{pH\ 5} \text{D-glucose} \rightleftharpoons [CHO{-}HCOH{-}HOCH{-}HCOH{-}HCOH{-}CH_2OH] \xrightarrow{pH\ 9} CH_2OH{-}HCOH{-}HOCH{-}HCOH{-}HCOH{-}CH_2OH$$

c. *Diborane.* Unlike the hydride ions, which are electron-rich nucleophiles, diborane is the dimer of an electron-deficient electrophilic species (BH_3). It is formed by reaction of a borohydride ion with strong acids and is used in 2,2′-dimethoxydiethyl ether (diglyme) as the solvent.

Diborane reacts faster with carboxylic acids, reducing them to primary alcohols, than with esters, because the carbonyl character of the adduct of the first is stabilized by resonance and that of the second is destabilized by resonance.

Diisoamylborane, which does not cleave ester groups (as does diborane), reduces acetylated aldono-1,4-lactones to the corresponding acetylated aldofuranoses.

$$2(BH_4)^- + 2H^+ \longrightarrow B_2H_6 + 2H_2$$

$$RC(=O)H + B_2H_6 \longrightarrow \left[R{-}C(=O){-}O{-}B\langle \;\updownarrow\; R{-}C(=O){-}\overset{+}{O}{=}\overset{-}{B}\langle \right] \longrightarrow RCH_2OH$$

$$RC(=O)OR + B_2H_6 \longrightarrow \left[R{-}C(=\overset{+}{O}{:}\overset{-}{B}H_3){-}OR' \;\updownarrow\; R{-}C(O{:}\overset{-}{B}H_3){=}OR' \right] \longrightarrow RCH_2OH + R'OH$$

d. *Sodium Amalgam.* Sodium amalgam in water and sodium in liquid ammonia were commonly used as reducing agents but have now been replaced by sodium borohydride. The metal alloy has an electron-rich surface that reacts with protons of a proton-donating solvent to form hydrogen atoms, which may combine to form molecular hydrogen or react with the substrate to give a radical ion. The same radical ion may be obtained directly from the metal, which in both cases is oxidized to a cation, giving up electrons to the substrate. The hydrogen radicals and the substrate radical ions, which are bound by adsorption to the electron-rich surface of the metal, combine to form the alkoxide ion, which then accepts a proton from the solution.

In the early literature, the reduction of lactones to aldoses (a key step in ascending the series) was carried out with sodium amalgam in acid solution (pH 3), using such buffers as sodium hydrogenoxalate and sulfuric

$$M\cdot + H^+ \rightarrow M^+ + H\cdot$$

$$H\cdot + \text{C}{=}\text{O} \rightarrow \cdot\text{C}{-}\text{O}^-$$

$$M\cdot + \text{C}{=}\text{O} \rightarrow \cdot\text{C}{-}\text{O}^- + M^+$$

$$H\cdot + \cdot\text{C}{-}\text{O}^- \rightarrow \text{HC}{-}\text{O}^-$$

acid and, later, ion-exchange resins. These buffers are needed because the reduction does not proceed in alkaline solution (formation of the sodium salt of the aldonic acid impedes reaction with the nucleophilic reductant). The free acids are not reduced, but esters can be. Thus, methyl D-arabinonate is converted into D-arabinose by sodium amalgam. The susceptibility of lactones to reduction and the inertness of the free acids have been exploited in the preparation of glycuronic acids by reduction of the monolactones of aldaric acids.

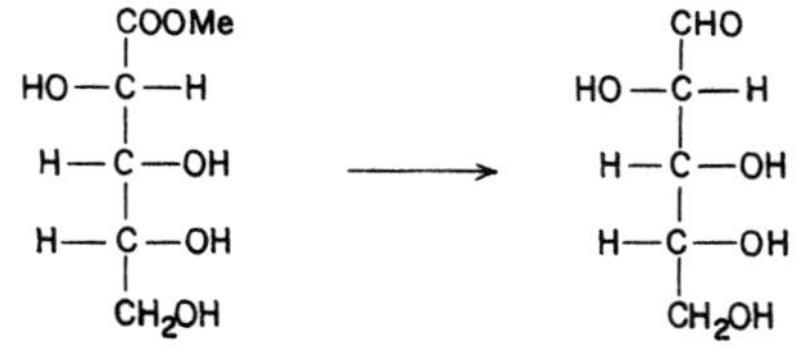

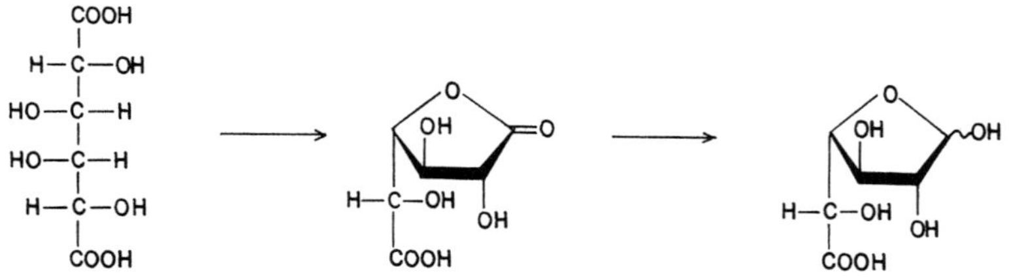

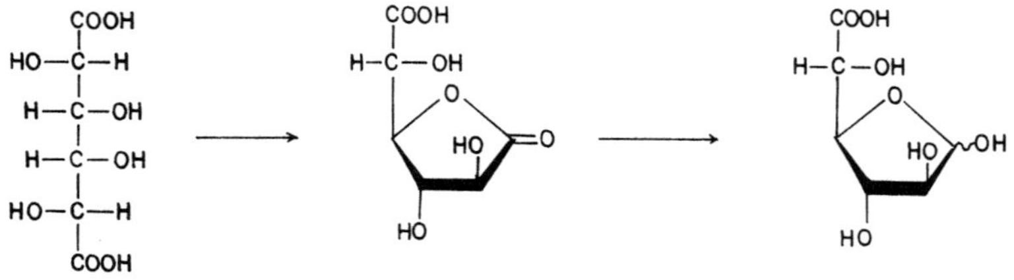

e. *Catalytic Hydrogenation*. In the course of hydrogenation with palladium or platinum catalysts, surface reactions occur between such electron-deficient species as the double bonds of carbonyl groups (which are attracted to the surface of the catalyst) and the hydrogen atoms generated by splitting hydrogen molecules.

Catalytic hydrogenation is used to convert keto groups into secondary alcohol groups. Thus, Raney nickel reacts with D-*xylo*-5-hexulosonic acid to give both D-gluconic and L-idonic acid. Usually, the axial alcohol is the preponderant product, because the reductant approaches the substrate from the less hindered side of the carbonyl group to form an equatorial carbon–hydrogen bond. This may be used to achieve an epimerization, as shown in the following example.

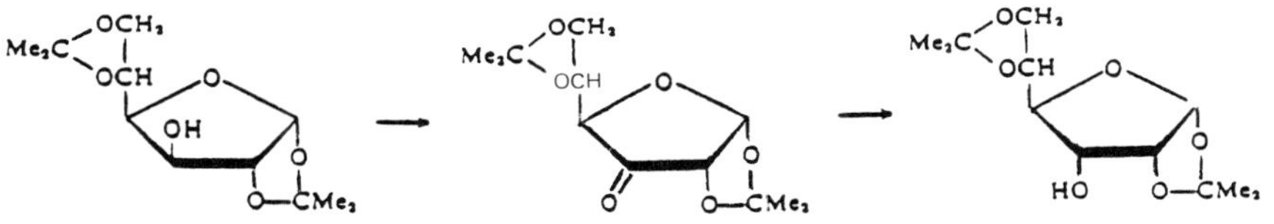

D-Gluco D-Allo

Reduction of carbonyl groups can be carried beyond the alcohol stage, to the deoxy stage. For example, methyl *β*-D-*ribo*-hexopyranosid-3-ulose is converted into the 3-deoxy derivative with hydrogen in the presence of platinum, and 1-deoxyalditols may be obtained from aldoses by reducing their dialkyl dithioacetals with Raney nickel.

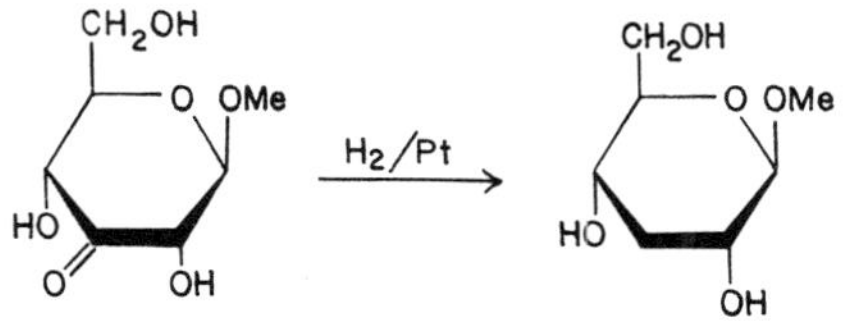

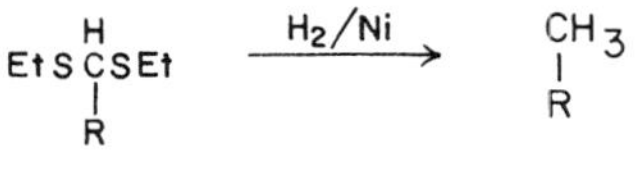

$$\text{EtS}\underset{\text{R}}{\overset{\text{H}}{\text{C}}}\text{SEt} \xrightarrow{H_2/Ni} \underset{\text{R}}{CH_3}$$

Reduction of a hydroxyl group to a deoxy function can be performed via the sequence tosyl → iodo → deoxy or, more directly by reducing the tosyl derivative in the case of primary tosyl groups. The latter reaction is, however, best performed with lithium aluminum hydride. An example of the preparation of a 6-deoxy derivative is the reduction of 6-deoxy-6-iodo-1,2:3,4-di-*O*-isopropylidene-α-D-galactopyranose with Raney nickel.

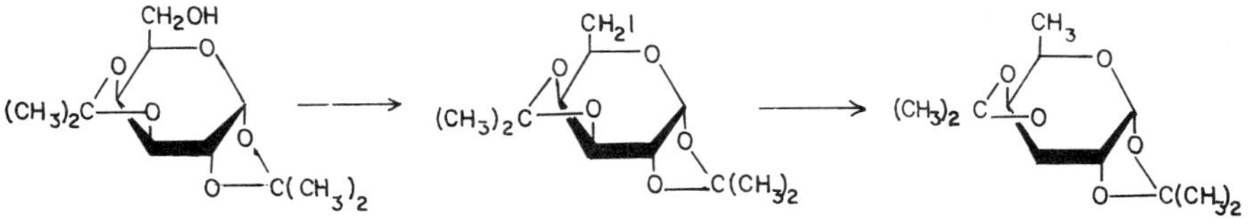

CH_2OH, $(CH_3)_2C$, $C(CH_3)_2$ → CH_2I, $(CH_3)_2C$, $C(CH_3)_2$ → CH_3, $(CH_3)_2C$, $C(CH_3)_2$

Another way to convert hydroxyl groups into methylene groups is by oxidizing the former to carbonyl groups and then reducing their hydrazones with borohydride. Alternatively, the conversion may be achieved via oxyphosphonium intermediates, as shown by Rydon.

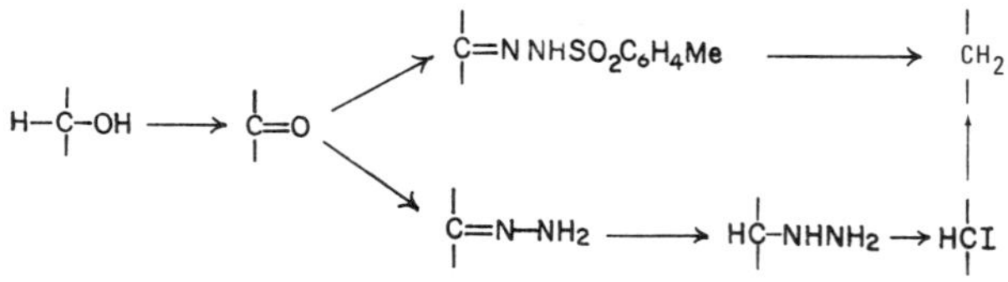

H—C—OH → C=O → C=N NHSO$_2$C$_6$H$_4$Me → CH$_2$

C=O → C=N—NH$_2$ → HC—NHNH$_2$ → HCI → CH$_2$

f. *Zinc Dust and Acetic Acid.* The reduction of glycopyranosyl halides with zinc dust and acetic acid produces 1,5-anhydrohex-1-enitols, commonly called glycals. The reaction is believed to proceed via an intermediate carbonium ion, which acquires a pair of electrons during the reduction step and loses an acetate ion. A closer look at the overall reaction reveals that the elimination of the electronegative groups could be achieved with bases and that a reducing agent is not necessary as was originally presumed.

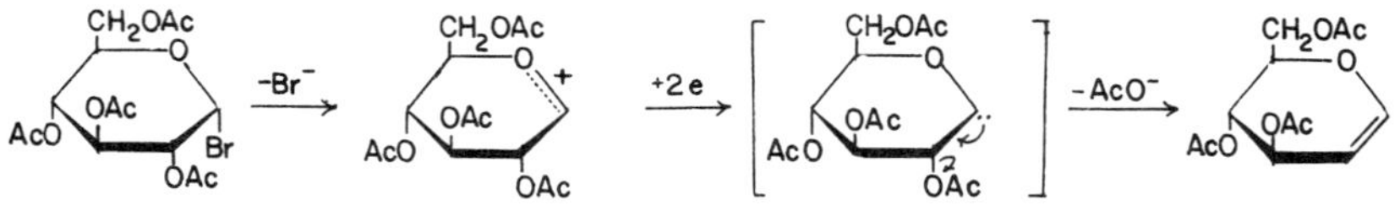

CH$_2$OAc, OAc, AcO, Br, OAc —Br^-→ CH$_2$OAc, OAc, AcO, OAc, + —+2e→ [CH$_2$OAc, OAc, AcO, OAc] —AcO^-→ CH$_2$OAc, OAc, AcO

Glycals are useful starting materials for the synthesis of deoxy sugars, as may be seen from the following sequence of reactions.

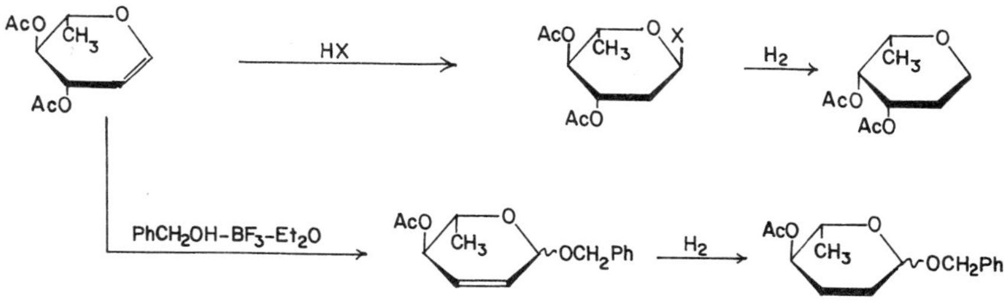

F. Planning Synthetic Schemes

It is recommended that multistep syntheses be planned by retrosynthetic analysis. The formula of the target compound is drawn first, followed by that of its possible precursors, and so on, until a synthon or a chiron (chirons are chiral synthons used to introduce sequences of chiral atoms in larger molecules) is reached that is accessible from commercially available starting materials. Retrosynthetic schemes are distinguished from actual schemes (those that begin with a starting material and end with a product) by the shape of the arrows. The arrows are broad and pointed toward the precursor in the former schemes and narrow and pointed toward the product in the latter ones.

Arrow in a hypothetical retrosynthetic scheme:

Product $\Rightarrow$ precursor

Arrow in an actual reaction sequence:

Precursor $\rightarrow$ product

1. Chiral Templates and the Chiron Approach to Synthesis

The term synthon is generally used to designate intermediates that could be used to construct complex molecules, for example, simple carbonium ions or carbanions. The term was originally used by E. J. Corey in connection with the retrosynthetic analysis of natural products. Hanessian coined the terms *chiron* to designate enantiomerically pure synthons, *chiral templates* to describe chiral synthons that can be used as templates to replicate segments of target molecules, and the *chiron approach* as a tool to be used in retrosynthetic analysis. The latter approach involves scrutinizing the molecular architecture of a target compound and engaging in a visual dialogue to locate elements of symmetry, chirality, and functionality that are found in accessible chiron, such as fragments of carbohydrate molecules. The best chirons are those that contain the largest number of chiral centers that can be relocated with the minimum perturbation. The approach has been computerized and the software is now available.

For many years chemists have considered carbohydrates as simple cyclic five- or six-membered ring compounds, substituted with hydroxyl groups (and occasionally with amino or carboxyl functions), that are amenable to peripheral transformations such as the introduction of amino and deoxy functions. Today, carbohydrates are also viewed as chirons useful in the synthesis of chiral natural products. In principle, three types of natural products can be constructed from carbohydrate chirons: (i) those that have apparent carbohydrate-type symmetry, for example, compounds that possess chiral tetrahydrofuran or tetrahydropyran structures, such as nucleosides; (ii) targets that have partially hidden carbohydrate-type symmetry, such as cyclic and acyclic compounds containing chiral appendages; and (iii) compounds having hidden carbohydrate-type symmetry, for example, alkaloids, macrolides, and prostaglandins. The carbohydrate portions in these compounds are difficult to discern because they constitute small segments of the molecule. To help detect possible carbohydrate chirons in complex molecules, Hanessian developed an empirical "rule of five." The target molecule is divided into segments containing four or five carbon atoms in such a way as to have at one end an sp^2 carbon atom (which will correspond to the anomeric carbon of a sugar) and at the other end an oxygen-bearing carbon atom (which will correspond to C-4 of a furanose or C-5 of a pyranose). In selecting possible chirons maximum overlap with the segments is sought, particularly in regard to the nature, location, and stereochemistry of the substituents. To exemplify the use of saccharide chirons in the synthesis of natural products, the retrosynthetic analysis of four target compounds will be discussed. For additional examples the reader is referred to Professor Hanessian's book *Total Synthesis of Natural Products; The Chiron Approach,* which contains several more examples.

2. Retrosynthetic Schemes Involving Carbohydrate Chirons

a. *The Amino Acid in Pyridomycin.* Inspection of the structure of this molecule indicates the presence of four different substituents on an acyclic five-carbon chain with a carboxyl terminus. It can also be seen that the stereochemistry of one of the hydroxyl-bearing carbon atoms could be protected by forming a furanose derivative. The heterocycle could be introduced by manipulating C-6 of 1,2-*O*-isopropylidene-α-D-glucofuranose; the C-methyl group could be introduced on a suitable ketone on C-3 via Wittig methodology, and a double inversion would afford the desired configuration of the amino group at C-5.

N HO NH2 CO2H Me ≡ Pyr NH2 HO CO2H Me ⟹ Pyr N3 O OH Me OH ⟹

RO N3 O O Me O Me Me ⟹ RO RO O O Me O Me Me ⟹ Me O Me O O O O O Me Me ⟹ D-glucose

b. *Erythronolide A*. It is possible to construct the two acyclic segments of the macrolide antibiotic, erythronolide A from chirons A and B, which share the same substitution pattern. It can be seen that starting with methyl α-D-glucopyranoside, it is possible to make use of the conformational bias and anomeric stereoselection provided by the axial orientation of the anomeric substituent to introduce the functional group in the desired configuration. [For the actual synthesis see S. Hanessian and G. Rancourt, *Can. J. Chem.* **55,** 1111 (1977).]

Me O Me Me HO OH OH Me Me OH Me OH O Me O OH Me Me CHO OH OH Me OH Me OH HO2C OH Me ⟹ OH Me O OH HO OMe Me Chiron B

O OR OH Me OMe Me Chiron A

⟹ RO Me OH O OMe Me ⟹ RO O OR' O OMe Me ⟹ RO O O OMe R'O Me invert ⟹ RO O HO OMe R'O Me oxidize

⟹ RO O R'O Me OMe O invert reduce ⟹ RO O R'O Me OMe OH ⟹ RO O R'O OMe O ⟹ RO O OH HO OMe OH

c. *Prostaglandin* $F_2\alpha$. This compound could conceivably be synthesized from D-glucose by chain extension, making use of the two central hydroxyl groups of the alditol intermediate to create a trans double bond and one unprotected hydroxyl group. This is a good example of the use of a Claisen rearrangement to create a center of chirality in the β position of an existing center. [For the actual synthesis, see G. Stork, T. Takahashi, I. Kawamoto, and T. Suzuki, *J. Am. Chem. Soc.* **100,** 8272 (1978).]

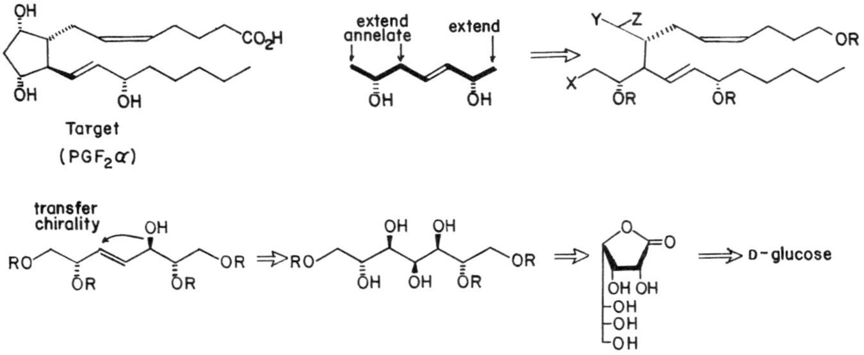

d. *Anisomycin.* The retrosynthetic analysis of this compound reveals that a carbohydrate ring structure may be used to construct segments of the target around it. For example, the pyrrolidine ring could be formed by intramolecular cyclization of a primary amine with inversion of configuration at a ring carbon atom. The original carbohydrate portion would then become an appendage or an extension of the new heterocyclic ring (see p. 165). The actual synthesis was carried out by Moffatt *et al.* [*Pure Appl. Chem.* **51,** 1363 (1978).]

In the following examples, the retrosynthetic analysis of some interesting target natural products will be followed by their actual synthesis.

1. *Negamycin:* This antibiotic contains an amino acid (2-hydroxy-β-lysine) attainable from 3,6-diamino-2,3,4,6-tetradeoxy-L-*threo*-hexonolactone. Inspection of the stereochemical features of the latter suggests that galactose could be used as a possible starting material. The actual synthesis started with D-galacturonic acid, which was esterified with methanol, acetylated, and converted into a glycal by treatment with HBr and then with zinc. The 1,2-double bond thus formed was subjected to an electrophile (I_2) mediated, nucleophilic attack with MeOH, to yield after reduction a methyl 2-deoxyglycoside, which was treated with base to cause a β elimination and form another double bond, this time in position

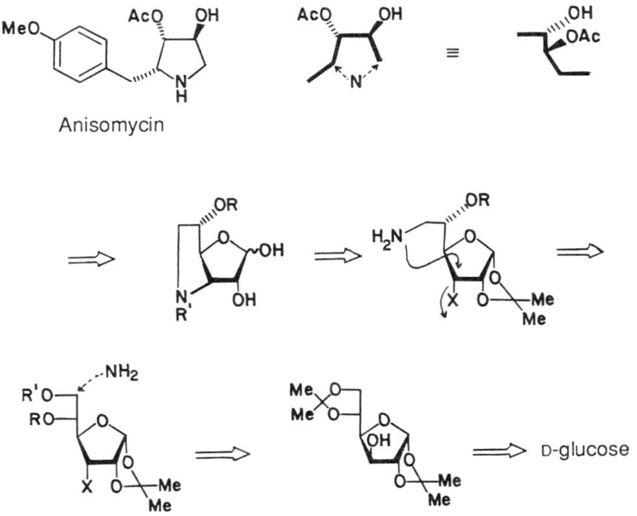

Anisomycin

D-glucose

4. The double bond was hydrogenated (Pd/C) to produce a deoxy function at C-4, with inversion of configuration at C-5. Reduction of the ester group (LAH) and mesylation of the free hydroxyl groups, followed by (a) treatment with sodium azide, (b) reduction, and (c) acetylation, introduced acetamido functions on C-3 and C-6. Finally, hydrolysis of the methyl glycoside and oxidation of C-1 with bromine gave the lactone of the desired amino acid [see S. Kondo, S. Shibahara, S. Takashiki, K. Maeda, and M. Ohno, *J. Am. Chem. Soc.* **93,** 6305 (1971)].

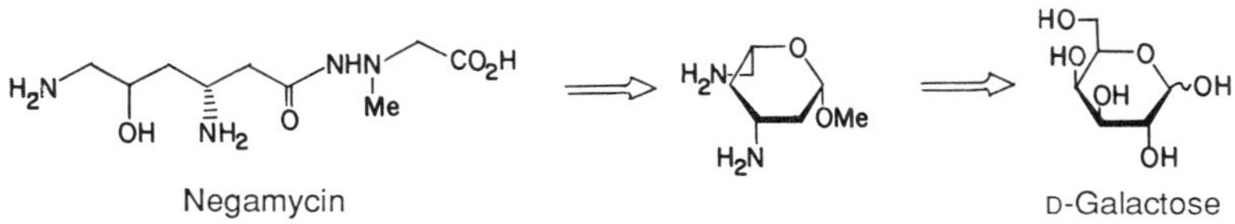

Negamycin

D-Galactose

2. *Oxybiotin and 11-Oxaprostaglandin* $F_2\alpha$: Examination of the structures of these oxygen analogs of biotin and of prostaglandin reveals that 1,2 : 5,6-di-*O*-isopropylidene-α-D-glucofuranose might serve as a possible chiron in their preparation. Actually, the first target compound was prepared from the dicyclohexylidene derivative, which was oxidized with RuO_4 to give a 3-ulose. Reduction with sodium borohydride followed by tosylation and treatment of the resulting ester with sodium azide yielded an azide having the D-*gluco* configuration. The 5,6-acetal protecting group was removed with acetic acid and the product oxidized with periodate to cleave the side chain and afford an aldehyde, which was reduced (sodium

R = CO_2Me

negamycin

[S. Shibahara, S. Kondo, K. Maeda, and H. Umezawa, *J. Am. Chem. Soc.* **94**, 4353 (1972).]

borohydride) and tosylated to yield 3-azido-1,2-*O*-cyclohexylidene-3-deoxy-5-*O*-tosyl-D-xylofuranose. Replacement of the 1,2-acetal protecting group with a 1,1-dimethyl acetal group (methanolic HCl) yielded 2,5-anhydro-3-azide-D-xylose-dimethyl acetal. This was subjected to another round of tosylation and treatment with azide, followed by reduction, benzamidation, and removal of the dimethyl acetal protecting group. To obtain the target compound, the resulting aldehyde was subjected to a Wittig reaction, hydrolyzed to remove the benzoyl groups, and treated with phosgene.

The second target (see scheme on p. 168) was obtained from a similar chiron, 1,2-*O*-isopropylidene-α-D-glucofuranos-3-ulose by a Wittig reaction (to introduce the acetic acid side chain). Treatment with acetic acid and then acetic anhydride resulted in deblocking and then acetylation. The resulting branched-chain allofuranose derivative was then treated with acid to remove the 1,2-isopropylidene group and form a γ-lactone. Conversion to a phenyl 1-thioglycoside followed by reduction with Raney nickel and deacetylation afforded a deoxygenated bicyclic lactone having a dihydroxyethyl side chain. The chain was split with periodate, giving an aldehyde, which was subjected to a Wittig reaction to introduce the first side chain. The other side chain was introduced by reducing the lactone to an aldehyde with DIBAL and carrying out another Wittig reaction. Removal of the THP group yielded the optically pure prostaglandins, epimeric at C-15.

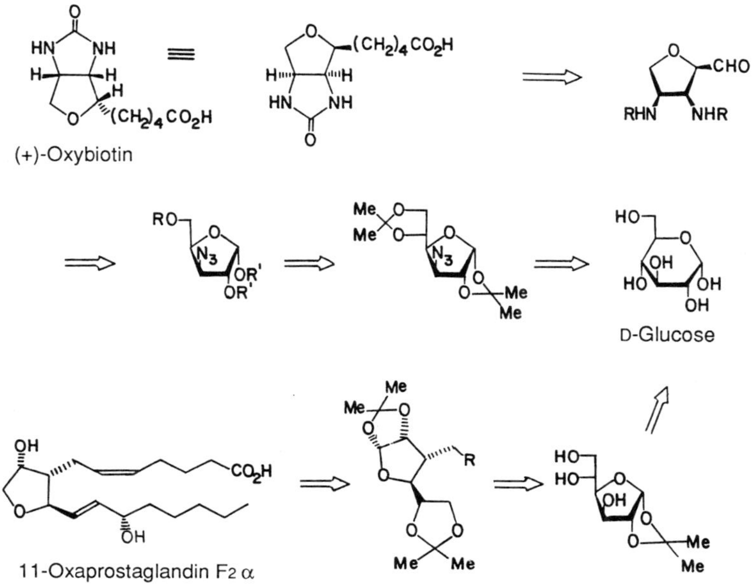

O
HN NH ≡ O $(CH_2)_4CO_2H$ ⟹ O CHO
H H H H
H HN NH RHN NHR
O $(CH_2)_4CO_2H$ O

(+)-Oxybiotin

⟹ RO O ⟹ Me O Me O O ⟹ HO O
N_3 OR' N_3 O Me OH
OR' O Me HO OH
OH

D-Glucose

Me
OH Me O
O
CO_2H ⟹ O R ⟹ HO O
HO
O OH
O O Me
OH O O Me
Me Me

11-Oxaprostaglandin F2 α

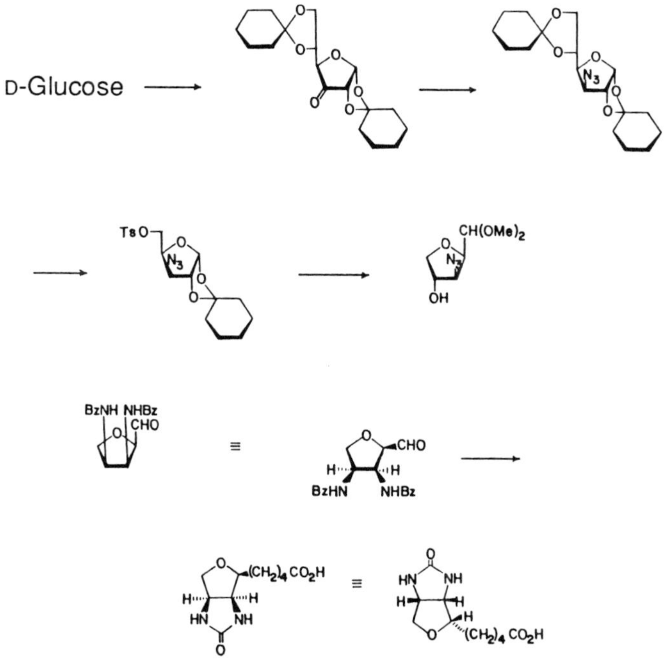

D-Glucose ⟶ O O ⟶ O O
O O O O
O N_3
O O
O O

⟶ TsO O ⟶ $CH(OMe)_2$
N_3 O N_3
O OH
O

BzNH NHBz
O CHO ≡ O CHO ⟶
H H
BzHN NHBz

O $(CH_2)_4CO_2H$ ≡ O
H H HN NH
HN NH H H
O H
O $(CH_2)_4CO_2H$

[H. Ohuri, H. Kuzuhara, and S. Emoto, *Agric. Biol. Chem.* **35**, 754 (1971).]

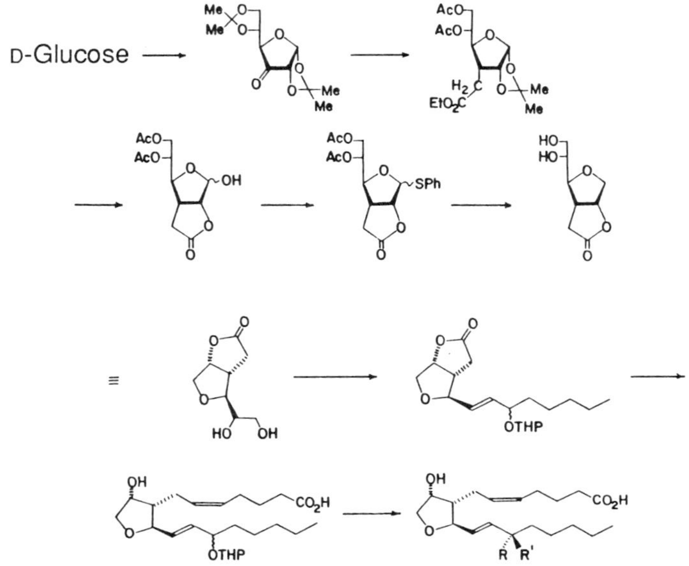

[G. J. Lourens and J. M. Koekemoer, *Tetrahedron Lett.* p. 3715 (1975).]

2. Accessible Carbohydrates and Carbohydrate Derivatives

It is essential for a chemist interested in preparing saccharide derivatives, or in using saccharides as chiral templates to introduce a sequence of chiral carbon atoms into a natural-product molecule, to be aware of the possible carbohydrates and carbohydrate derivatives that can be used as starting materials or chirons. Table III shows commercially available monosaccharides, and Table IV lists commercially available monosaccharide derivatives that can be used as starting materials in organic synthesis. Although other saccharides can be purchased, their use is usually restricted (because of high price) to analytical work, for example, as reference compounds for chromatography. If a saccharide is selected as a starting material, it must first be converted to the desired form. For example, if the synthon needed is acyclic, the saccharide must be converted to an acyclic derivative by conversion to a dithioacetal, an acyclic hydrazone, or a bis(hydrazone). If a pyran derivative is desired, the peracetylated sugar can be used, since it occurs in the pyranose form; whereas if a five-membered (furan) derivative is required, a methyl furanoside or 1,2 : 5,6-di-*O*-isopropylideneglycofuranose may be envisaged. Useful carbohydrate intermediates are discussed in the next section, followed by a

TABLE III

Commercially Available Monosaccharides[a]

Aldohexoses	D-Glucose
	D-Mannose
	D-Galactose
Aldopentoses	D-Ribose
	D- and L-Arabinose
	D-Xylose
Ketohexoses	D-Fructose
	L-Sorbose
6-Deoxy sugars	D- and L-Fucose
	L-Rhamnose

[a] Listed here are the monosaccharides commonly used in synthesis. Other monosaccharides are available for analytical use, for example, as reference compounds for gas chromatographic or liquid chromatographic analysis.

discussion of protecting and leaving groups useful in saccharide synthesis.

3. Useful Intermediates in Synthesis

In planning a synthetic scheme, the commercial availability of saccharides and saccharide derivatives (see Tables III and IV), as well as their accessibility through synthesis, should be taken into consideration. The following are some useful intermediates that are either commercially available or readily accessible by synthesis from commercially available products:

a. *Peracetylated Sugars.* Penta-*O*-acetyl-α- and -β-D-glucopyranose, α-D-mannopyranose, and α- and β-D-galactopyranose are commercially available. In addition, tetra-*O*-acetyl-D-ribopyranose and -furanose are available in the β-D form. Despite the availability of these peracetylated sugars, most synthetic laboratories tend to prepare large quantities of the peracetylated sugars for use when needed. Peracetylated sugars are used mainly to provide access to the glycosyl halides, which in turn may be used to prepare glycosides and nucleosides.

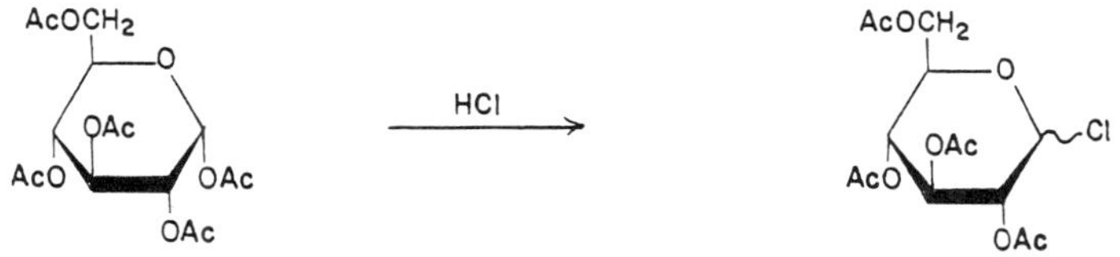

TABLE IV

Commercially Available Monosaccharide Derivatives

Cyclic derivative	Alditols, acids, and lactones
Pentoses	
D- and L-Arabinose	
Tri-*O*-benzyl-D-arabinofuranose (α and β)	
D-Ribose	Ribitol
2-Deoxy-D-*erythro*-pentose	
Tetra-*O*-acetyl-D-ribopyranose (β)	
Tetra-*O*-acetyl-D-ribofuranose (β)	
1-*O*-Acetyltri-*O*-benzoyl-D-ribofuranose	
D-Lyxose	
D-Xylose	
1,2-*O*-Isopropylidene-α-D-xylofuranose	
1,2:3,5-Di-*O*-isopropylidene-α-D-xylofuranose	
Hexoses	
D-Glucose	D-Glucitol, D-glucuronic acid, and 1,4-lactone
Penta-*O*-acetyl-D-glucopyranose (α and β)	D-Gluconic acid and 1,4-lactone
Methyl D-glucopyranoside (α and β)	D-Glucaric acid
Tri-*O*-acetyl-D-glucal	
1,2:5,6-Di-*O*-isopropylidene-α-D-glucofuranose	
1,2-*O*-Isopropylidene-α-D-glucofuranose	
D-Glucosamine · HCl	D-Glucosaminic acid
N-Acetyl-D-glucosamine	
2-Deoxy-D-*arabino*-hexose	
D-Mannose	D-Mannitol, D-manuronic acid, and 1,4-lactone
Methyl α-D-mannopyranoside	
Mannosamine · HCl	
N-Acetylmannosamine	
6-Deoxy-D-mannose (L-rhamnose)	
D-Galactose	D-Galactitol, galacturonic acid, 1,4-lactone, D-Galactaric acid
Penta-*O*-acetyl-D-galactopyranose (α and β)	
Methyl D-galactopyranoside (α and β)	
D-Galactosamine · HCl	
N-Acetyl-D-galactosamine	
2-Deoxy-D-*lyxo*-hexopyranose	
6-Deoxy-D- and -L-galactopyranose (D- and L-fucose)	
D-Fructose	D-Glucoascorbic acid
L-Sorbose	L-Ascorbic acid
Heptoses	
D-*glycero*-D-*gluco*-Heptose	D-*glycero*-D-*gluco*-Heptono-1,4-lactone

Haloacetates are more labile than acetates and are useful temporary blocking groups for the synthesis of disaccharides. For example, an *O*-bromoacetyl group can be selectively hydrolyzed with thiourea, which does not affect acetyl groups. The latter require treatment with ethanolic NaOEt or NH_3 for hydrolysis.

b. *Methyl Glycosides.* The methyl glycopyranosides of hexoses are readily available in crystalline form, whereas the methyl furanosides are usually glasses that are converted into crystalline derivatives and stored as such. Methyl α- and β-D-glucopyranoside and methyl α- and β-D-galactopyranoside, as well as methyl α-D-mannopyranoside, are commercially available. In addition, the methyl glycofuranosides of these hexoses and the glycopyranosides and glycofuranosides of aldopentoses are readily accessible by synthesis.

Methyl glycosides may be converted into glycosyl halides by acetylation followed by treatment with HBr in acetic acid, or by acetylation and acetolysis (treatment with Ac_2O and H_2SO_4) followed by reaction with HBr in dichloromethane.

AcOCH2, O, OMe, AcO, OAc — Ac_2O/H_2SO_4 → AcOCH2, O, OAc, AcO, OAc; HBr → AcOCH2, O, Br, AcO, OAc; ROH → AcOCH2, O, OR, AcO, OAc

c. *Glycosyl Halides.* Glycosyl halides are excellent intermediates because their halogen atoms can be readily replaced by nucleophiles. Because their shelf life is short, few are commercially available, and they must be used soon after preparation. Glycosyl halides can be obtained from peracetylated methyl pyranosides or furanosides by replacing the OMe group with HCl or HBr in acetic acid or, more readily, by replacing the OAc group of 1-acetylated sugars with HCl or HBr in dichloromethane. Glycosyl halides are used to provide access to such 1-substituted sugar derivatives as glycosides and nucleosides, which are obtained by nucleophilic substitution using an alcohol or an amine, respectively, in the presence of a halogen acceptor. In addition, the glycopyranosyl halides are used to give access to glycals (furanosyl halides are not well suited for this reaction, because their glycals are extremely labile and aromatize to furan derivatives). Glycosyl halides have been used in Friedel–Crafts reactions (electrophilic substitution) with activated aromatic rings.

d. *Glycals.* The acylated glycals are prepared from per-*O*-acetylglycopyranosyl halides by reduction with zinc dust and acetic acid. These 1,2-unsaturated pyranoid derivatives constitute excellent starting materials for 2-deoxypyranose derivatives.

AcOCH2, O, OAc, AcO, Cl, OAc —Zn AcOH→ AcOCH2, O, OAc, AcO

The latter are obtained by treating the glycal with HCl and then treating the resulting 2-deoxyglycosyl halide with a nucleophile. Alternatively, they can be used to prepare 2,3-dideoxypyranoses by treating the glycal with an alcohol in the presence of a Lewis acid.

AcOCH2, O, OH, AcO —HCl→ AcOCH2, O, OAc, AcO, Cl —OR→ 2-deoxyglycosides

↓ ROH Lewis acid

AcOCH2, O, OR, AcO —H2→ AcOCH2, O, OR, AcO

e. *Epoxides.* The hydroxyl groups on C-2 and C-3 of a xylofuranose or of an altropyranose are in trans-diaxial positions relative to one another, and this renders them susceptible to epoxide ring formation. Hydrogenation of a 2,3-epoxide attached to a furanose ring yields preferentially the 3-deoxy isomers, whereas in pyranose rings, the isomer having an axial hydroxyl group is favored. This last rule was first observed by A. Fürst and P. Plattner in steroids.

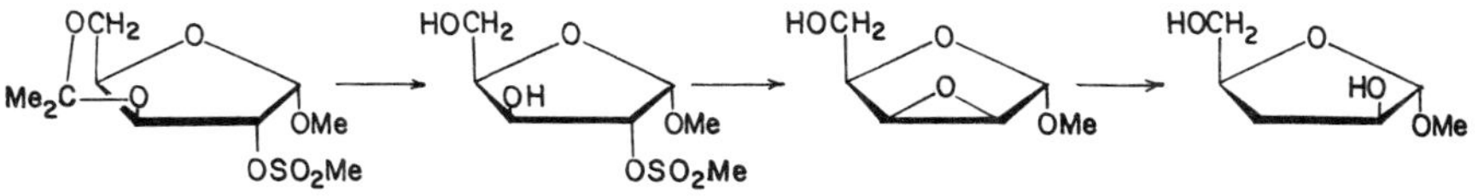

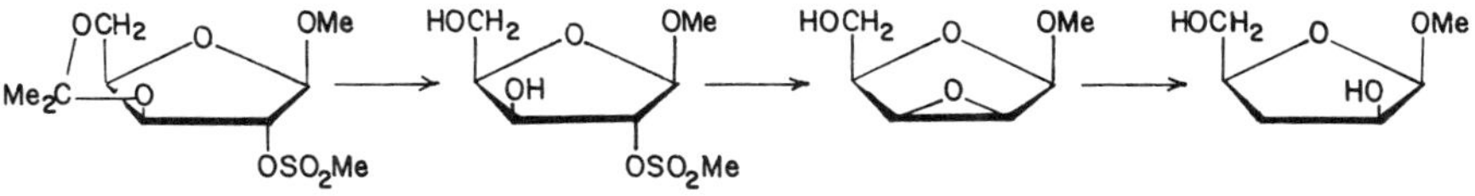

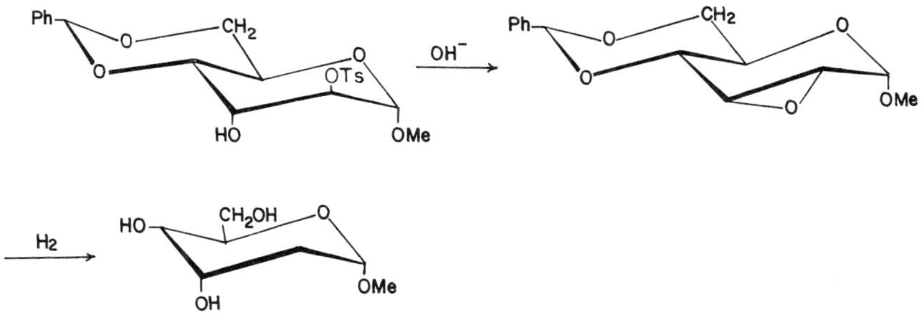

f. *Isopropylidene Derivatives.* Isopropylidene acetals are valuable starting materials for gaining access to furanosides from crystalline materials. Furthermore, as these derivatives block two hydroxyl groups at a time, they can be used to direct an entering group into the vacant location(s). For example, the di-*O*-isopropylidene derivatives of hexofuranoses and the mono-*O*-isopropylidene derivative of methyl pentofuranosides have only one hydroxyl group free, the monoisopropylidene acetals of hexofuranoses have three free hydroxyl groups, and the monoisopropylidene acetals of pentofuranoses have two such free groups. The following are some of the useful isopropylidene acetals.

1. 1,2:5,6-Di-*O*-isopropylidene-α-D-glucofuranose is commercially available, or it may be obtained in one step from D-glucose by treatment with acetone and an acid catalyst. It is used to give access to the O-3 of an aldohexose or aldopentose; this is achieved either by nucleophilic substitution of a leaving group or introduced onto O-3, or by oxidation of the 3-hydroxyl group to a keto group and then derivatizing the ketone. The diisopropylidene acetal of D-glucose, like all the diisopropylidene acetals

of other sugars, may be partially hydrolyzed to give the 1,2-isopropylidene acetal. The latter derivative can be oxidized with periodate and the product reduced with borohydride to give a pentofuranose derivative. The di-*O*-isopropylidene derivatives of glucose and mannose were used by Fraser-Reid and co-workers to synthesize an antifungal lactone, canadensolide [see R. C. Anderson and B. Fraser-Reid, *J.C.S. Chem. Commun.* p. 556 (1980)], and a sesquiterpene, germacronolide [see T. F. Tam and B. Fraser-Reid, *J. Org. Chem.* **45,** 1344 (1980)].

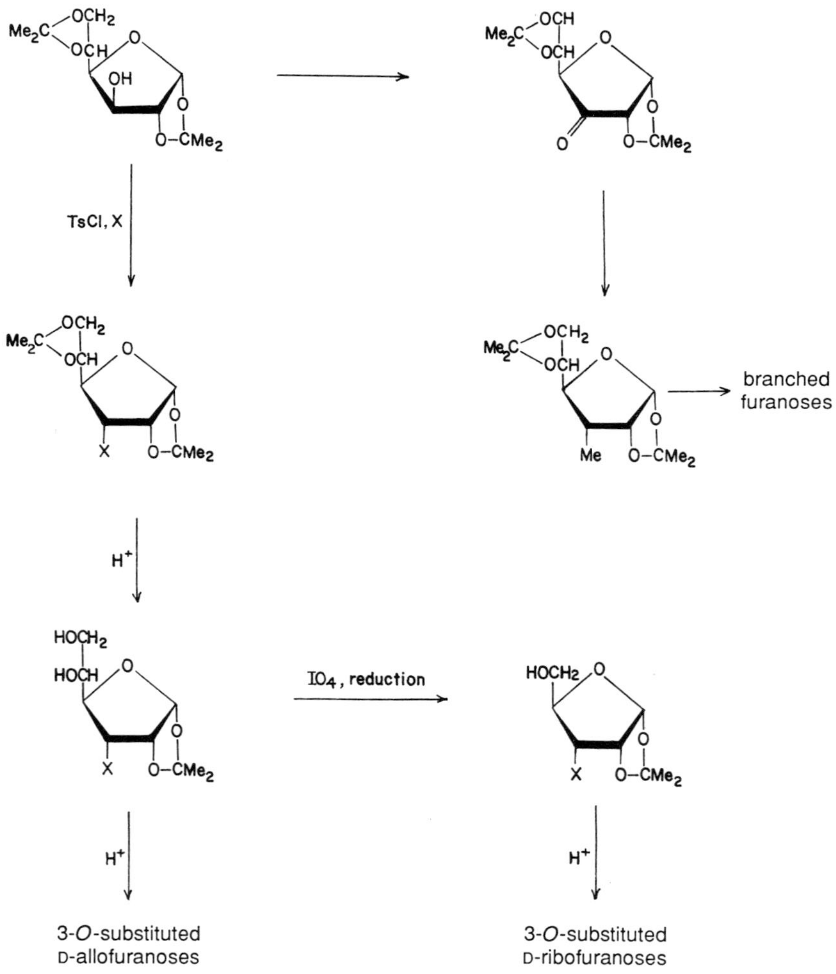

Me$_2$C, OCH$_2$, OCH, O, OH, O, O–CMe$_2$

Me$_2$C, OCH, OCH, O, O, O, O–CMe$_2$

TsCl, X

Me$_2$C, OCH$_2$, OCH, O, O, X, O–CMe$_2$

Me$_2$C, OCH$_2$, OCH, O, O, Me, O–CMe$_2$

branched furanoses

H$^+$

HOCH$_2$, HOCH, O, O, X, O–CMe$_2$

IO$_4$, reduction

HOCH$_2$, O, O, X, O–CMe$_2$

H$^+$

H$^+$

3-*O*-substituted D-allofuranoses

3-*O*-substituted D-ribofuranoses

2. 1,2:3,5-Di-*O*-isopropylidene-α-D-xylofuranose and the 1,2-*O*-isopropylidene derivative of this sugar are commercially available. The latter compound has two free hydroxyl groups, which can be blocked by different blocking groups by making use of the fact that one is a primary and the other is a secondary hydroxyl group. Another useful derivative of D-xylose is obtained by acetonating methyl α-D-xylofuranoside; the resulting methyl 3,5-*O*-isopropylidene-α-D-xylofuranoside has only one position free for reaction.

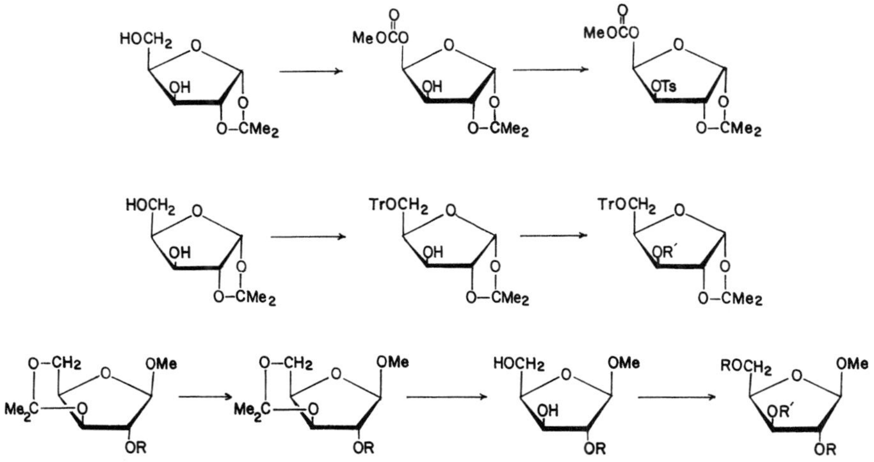

3. Methyl 2,3:5,6-di-*O*-isopropylidene-α-D-mannofuranose is readily obtained from D-mannose; it gives access to furanose acetals of D-mannose and D-lyxose.

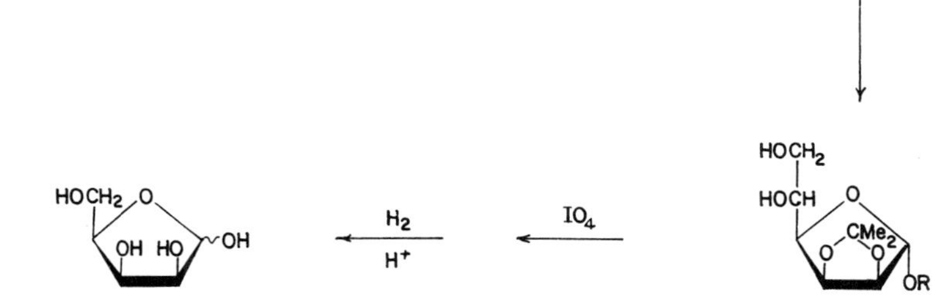

4. Methyl 2,3-*O*-isopropylidene-β-D-ribofuranoside, obtained in two steps from the free sugar, can be used to gain access to O-5 of D-ribose.

g. *Benzylidene Acetals.* The benzylidene acetals of methyl glycopyranosides are not commercially available but are readily prepared. They are used in the same way as the isopropylidene acetals, to block pairs of groups and subsequently leave two free hydroxyl groups for reaction. They are also used to introduce a halogen atom on the carbon atom of the previous primary hydroxyl group by using Hanessian's method employing NBS. The di-*O*-benzylidene derivatives have been used to introduce a 2-deoxy-3-keto functionality by reaction with BuLi; this reaction first gives the 2-en-3-ol, which tautomerizes.

The benzylidene acetals most often used are methyl 4,6-*O*-benzylidene-α-D-glucopyranoside and methyl 2,3:4,6-di-*O*-benzylidene-α-D-mannopyranoside. The latter was used by Horton and co-workers in the synthesis of the amino sugar daunosamine, as shown on p. 177 [see also D. Horton and W. Weckerle, *Carbohydr. Res.* **44,** 227 (1975)].

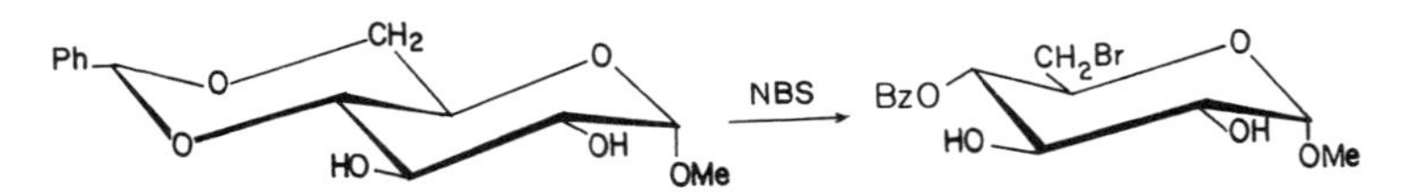

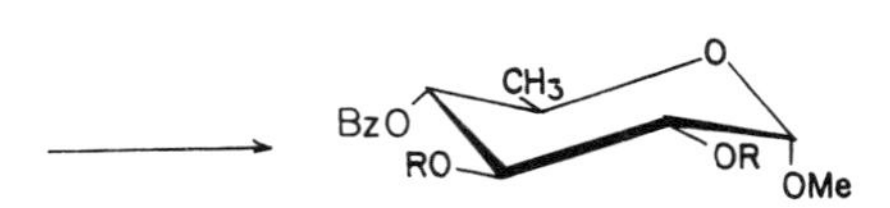

h. *Lactones.* Lactones are useful intermediates in chain extension reactions (ascending the series). Thus, in the Kiliani–Fischer synthesis, higher lactones are formed by reacting the aldehydo group of saccharides with cyanide and hydrolyzing the resulting nitrile. Attempts have been made to use lactones as C-nucleoside precursors. Blocked pentono-γ-lactones were reduced and subjected to Wittig reactions to yield hex-1-enitols, which underwent electrophile-induced cyclizations to produce ribofuranosyl derivatives of alkyl halides. For example, when the lactone group of 5-*O*-benzyl-2,3-*O*-isopropylidene-D-ribono-1,4-lactone was re-

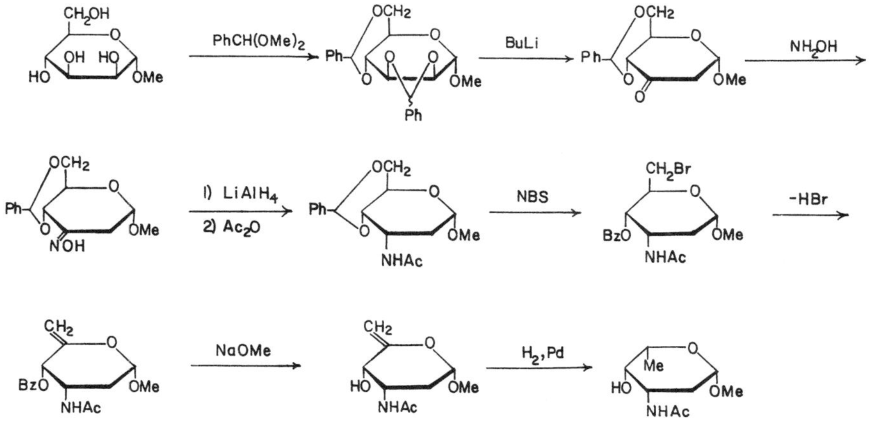

Horton's daunosamine synthesis. [See D. Horton and W. Weckerle, *Carbohydrate Research* **44,** 227 (1975).]

duced to a lactol (a hemiacetal) with DIBAL and the resulting aldehyde treated with a Wittig reagent ($Ph_3P{=}CH_2$), the acyclic 6-*O*-benzyl-1,2-dideoxy-3,4-*O*-isopropylidene-D-*ribo*-hex-1-enitol was produced. Electrophile (bromine or iodine) mediated cyclization of this compound led to addition of an electrophile on the unsubstituted terminus of the double bond and of oxygen on the other terminus. The preponderant product had the altro and not the allo configuration needed for nucleosides.

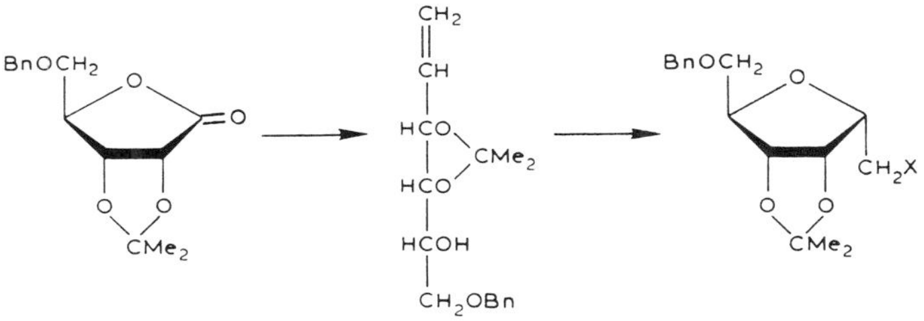

4. Useful Protecting and Leaving Groups

Discussed in the previous section are starting materials available for synthetic work that can be purchased or are readily obtained from commercially available saccharides. The design of a multistage synthesis involves the conversion of these synthons, via a number of intermediates, into desired products. In the chemical manipulations that follow, different functional groups in the molecule must be selectively protected. In this process some groups will be purposely left unprotected, to undergo subsequently a given reaction while the protected groups remain unaffected. The sequence in which the different reactions are carried out is of the greatest importance to ensure that the reagents used in one set of reactions are compatible with the groups present at the time. Furthermore, during the synthesis it is often necessary to block or deblock selectively some of the groups, and a notion of the reactivity of these groups becomes valuable. For these reasons, the synthetic chemist must be fully aware of the protecting groups available, as well as the conditions needed to introduce these groups (to ensure that the formation of one group does not cause the cleavage of another) and the conditions needed to cleave each of them (to ensure that elimination of a group does not result in splitting of a needed blocking group).

Table V shows the blocking groups commonly used in carbohydrate chemistry, arranged according to their types. These start with ethers, which are formed in acidic or basic media and are usually cleaved by acids. Benzyl and trityl ethers are, in addition, cleaved by catalytic hydrogenolysis. The ethers are followed by esters, which are usually formed by reaction with acid anhydrides or chlorides in pyridine and are cleaved by bases. Amides are formed in the same way but are best hydrolyzed by acids; otherwise, stronger bases are needed. Cyclic acetals are formed in the presence of acids and cleaved by strong acids.

In planning a synthesis using these blocking groups, care must be taken to ensure that cleavage of one group does not interfere with another, i.e., that the groups are compatible.

Table VI shows some leaving groups that are used to introduce substituents with inversion of configuration. The various sulfonyl groups and halogens are presented.

It should be noted that Tables V and VI do not list all of the known blocking and leaving groups. For a complete list of these, the reader is referred to an excellent text [T. W. Greene, "Protective Groups in Organic Synthesis." Wiley, New York, 1981].

5. Synthesis of Carbohydrates from Noncarbohydrates

The synthesis of saccharides from nonsaccharide precursors has attracted the attention of many researchers in the past decade. The most successful syntheses were those of saccharides having few chiral centers, such as tetroses, pentoses, and deoxyhexoses [for a review on this subject, see A. Zamojski, A. Banaszek, and G. Grynkiewicz, *Adv. Carbohydr. Chem. Biochem.* **40,** 1 (1982)].

a. *Synthesis of Tetroses.* When subjected to *cis*-hydroxylation (with silver chlorate, potassium permanganate, or peroxybenzoic acid), *cis*-

TABLE V
Blocking Groups

Ethers

1. Methyl ethers
   Formation: Methyl ethers are best prepared from a glycoside rather than from the free sugar. The following methylating agents have been used: (a) Me_2SO_4 (NaOH), (b) MeI (KOH/Me_2SO), (c) CF_3SO_2OMe (CH_2Cl_2/Py), (d) MeI (NaH/DMF).
   Cleavage:

   $ROMe + Me_3SiI \longrightarrow ROSiMe_3 + H_2O \longrightarrow ROH$

2. Methoxymethyl ethers (MOM ethers)
   Formation:

   $ROH + CH_2\,(OMe)_2 \longrightarrow RO{-}CH_2OCH_3$

   Cleavage:

   $ROCH_2OCH_3 + HCl \longrightarrow ROH$

3. 2-Methoxyethoxymethyl ethers (MEM ethers)
   Formation:

   $ROH + CH_3OCH_2CH_2OCH_2Cl \longrightarrow ROCH_2OCH_2CH_2OCH_3$

   Cleavage:

   $RO{-}MEM + TiCl_4/CH_2Cl_2 \longrightarrow ROH$

4. Tetrahydropyranyl ethers (THP ethers)
   Formation:

   ROH + dihydropyran (TsOH/CH_2Cl_2) $\longrightarrow$ RO—THP

   Cleavage: Mild acid conditions
5. Allyl ethers
   Allyl ethers are moderately stable to acidic conditions.
   Formation:

   $ROH + H_2C{=}CHCH_2Br \longrightarrow ROCH_2CH{=}CH_2$

   Cleavage:

   $ROCH_2CH{=}CH_2 + BuOK \longrightarrow ROCH{=}CHCH_3 + H^+ \longrightarrow ROH$

6. Benzyl ethers
   Formation:

   $ROH + PhCH_2Cl$ + NaH or $NaOH/Me_2SO \longrightarrow ROCH_2Ph$

   Cleavage: Catalytic hydrogenolysis (Pd/C)
7. Triphenylmethyl ethers (selective protection of primary OH)
   Formation:

   $RCH_2OH + PH_3CCl/C_5H_5N \longrightarrow ROCH_2OTr$

   Cleavage: Mild acid hydrolysis or catalytic hydrogenolysis

(continued)

TABLE V ***(continued)***

8. Trimethylsilyl (Me_3Si) ethers and *tert*-butyldimethyl- or diphenyl-silyl (Me_2-*t*-Bu or Ph_2-*t*-Bu) ethers
   Formation: By treating monosaccharides with chlorotrimethylsilane or hexamethyldisilazane in pyridine for the first and *tert*-butyldimethylchloro or chlorodiphenylsilane for the second
   Cleavage:

$$ROSiMe_3 + K_2CO_3/\text{methanol} \longrightarrow ROH$$
$$ROS:Me_2Ph + F^- \longrightarrow ROH$$

Esters

1. Acetates
   Formation: Ac_2O/pyridine
   Cleavage: KOEt or K_2CO_3 or NH_3, all in EtOH
2. Benzoates and *p*-substituted benzoates
   Formation: BzCl/pyridine
   Cleavage: Same as acetates, but more vigorous conditions needed

Amides (for amino sugars)

1. *N*-Acetyl derivatives
   Formation: Same as *O*-acetyl derivatives
   Cleavage: *N*-Acetyl groups are cleaved with acids, whereas *O*-acetyl groups are cleaved with buses. Thus, if *O*- and *N*-acetyl groups are present $Et_2O \cdot BF_3$ will remove only the *N*-acetyl groups, but refluxing with 1 *M* HCl removes both groups. *N*-Acetyl groups are less susceptible to base than are esters, and they require heating with hydrazine or $Ba(OH)_2$.
2. *N*-Trifluoroacetyl derivatives: Same as *N*-acetyl groups.

Cyclic Acetals

1. Benzylidene derivatives
   Formation: Benaldehyde/$ZnCl_2$; methyl glycopyranosides yield 4,6-benzylidene acetals.
   Cleavage:
   (a) Acid hydrolysis, (b) catalytic hydrogenolysis (Pd/C),
   (c) *N*-bromosuccinimide (NBS) yields the 4-*O*-benzoyl-6-bromo derivative.
2. Isopropylidene derivatives
   Formation: Acetone/acid or $Me_2C(OMe)_2$/TsOH. Methyl glycopyranosides yield six-membered 4,6-acetals under drastic conditions, while free sugars (e.g., D-glucose and D-mannose) give five-membered acetals.
   Cleavage: Acid media (warm HCl, TsOH, AcOH, Dowex 50-W). Selective hydrolysis of 1,2:5,6-diacetals yields 1,2-acetals.

TABLE VI

Leaving Groups

1. Alkane- and arene-sulfonyl groups
   (Reactivity: trifluoromethyl > methyl > *p*-tolylsulfonyl (tosyl) groups)
   Formation: Alcohol + sulfonyl chloride/pyridine
2. Halogens
   (Reactivity: I > Br; Cl generally less reactive)
   Formation:
   Alkane- or arene-sulfonate + NaI or NaBr
   Alkoxyphosphonium salts + halogens or halides

alkenes afford *erythro*-diols and *trans*-alkenes yield *threo*-diols. Stereochemically pure alkenes may be obtained by regioselective reduction of alkynes (the *cis*-alkenes by hydrogenation in the presence of Lindlar catalyst and the *trans*-alkenes by reduction with lithium aluminum hydride), or they may be available in one or both forms. For example, *trans*-2-butenoic (crotonic) acid, which is commercially available, can be converted to 4-deoxy-DL-threonic acid by treatment with peroxybenzoic acid. Using the same principle, it is possible to synthesize DL-threose from crotonaldehyde; the 1,1-diacetoxy acetal is brominated with *N*-bromosuccinimide, and the resulting 4-bromo derivative is treated with silver acetate and then subjected to *cis*-hydroxylation.

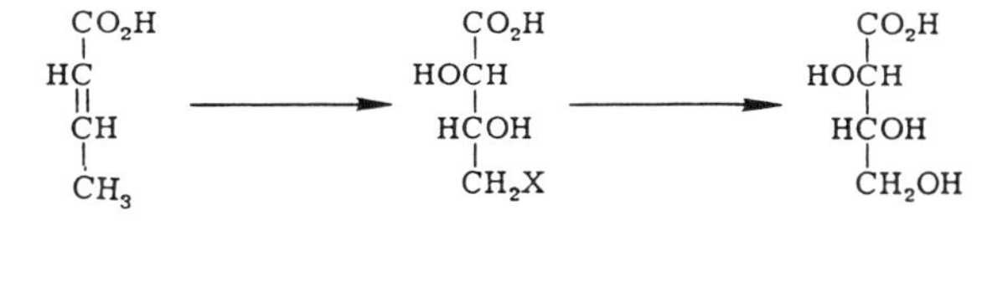

$CH(OAc)_2$ — CH = HC — CH_3 ⟶ $CH(OAc)_2$ — CH = HC — CH_2R ⟶ DL-Threose

R = Br
R = OAc

b. *Preparation of Pentoses, 2-Deoxypentoses, and Nucleosides*

(i) FROM PENTYNES AND PENTENES. 1,1-Diethoxy-5-(tetrahydropyran-2-yloxy)-3-pentyn-2-ol, obtained by adding 1-propynyl-3-(tetrahydropyran-2-yloxy) magnesium bromide to glyoxal monoacetal, can be hydrogenated over Lindlar catalyst or reduced with lithium aluminum hydride. The resulting *cis*- or *trans*-alkenes are then *cis*-hydroxylated to give DL-ribose and arabinose acetals (from the *cis*-alkene) and DL-xylose and lyxose acetals (from the *trans*-alkene), as shown on p. 183.

2-Deoxy-pentoses can be prepared from 1-methoxy-1-buten-3-yne by reaction with formaldehyde and methanol to give 5,5-dimethoxy-2-pentyn-1-ol. Hydrogenation of the triple bond in the presence of Lindlar catalyst, followed by *cis*-hydroxylation and hydrolysis, gives 2-deoxy-DL-*erythro*-pentose (2-deoxy-D-ribose), as shown on p. 183.

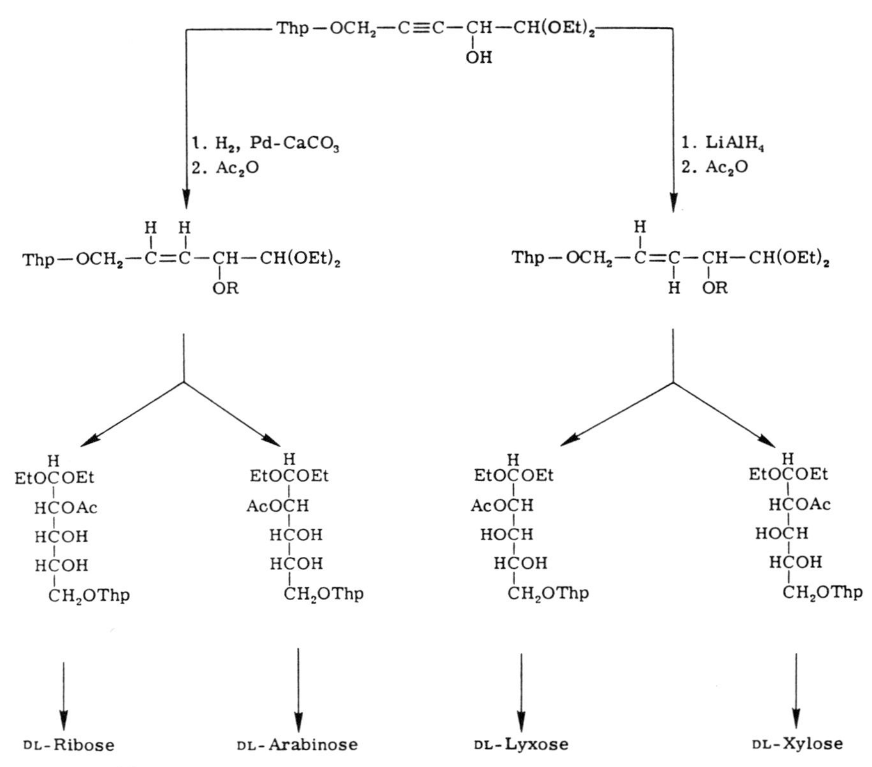

[T. Iwashige, *Chem. Pharm. Bull.* **9**, 492 (1961); I. Iwai and K. Tomita, *Chem. Pharm. Bull.* **11**, 184 (1963).]

(ii) FROM FURANS AND 2,3-DIHYDROFURANS. It is possible to achieve the stereocontrolled synthesis of C-nucleosides by constructing the ribose skeleton from a polybromoketone–iron carbonyl reaction. The starting material, a bicyclic ketone, is first prepared by cycloaddition of 1,1,3,3-tetrabromo-2-propanone with furan, using diiron nonacarbonyl, followed by reduction with a zinc–copper couple. Hydroxylation with catalytic amounts of osmium tetroxide, followed by reaction with acetone, gives an acetal, which is oxidized with trifluoroperoxyacetic acid to a lactone, from which the dimethylaminomethylene lactone (an intermediate in the synthesis of natural pyrimidine C-nucleosides) can be obtained, as shown on p. 183.

c. *Preparation of Branched-Chain Sugars.* The synthesis of the branched-chain sugar apiose was accomplished by partial hydrogenation of the benzylidene acetal of 4-ethoxy-2-(hydroxymethyl)-3-butyne-1,2-

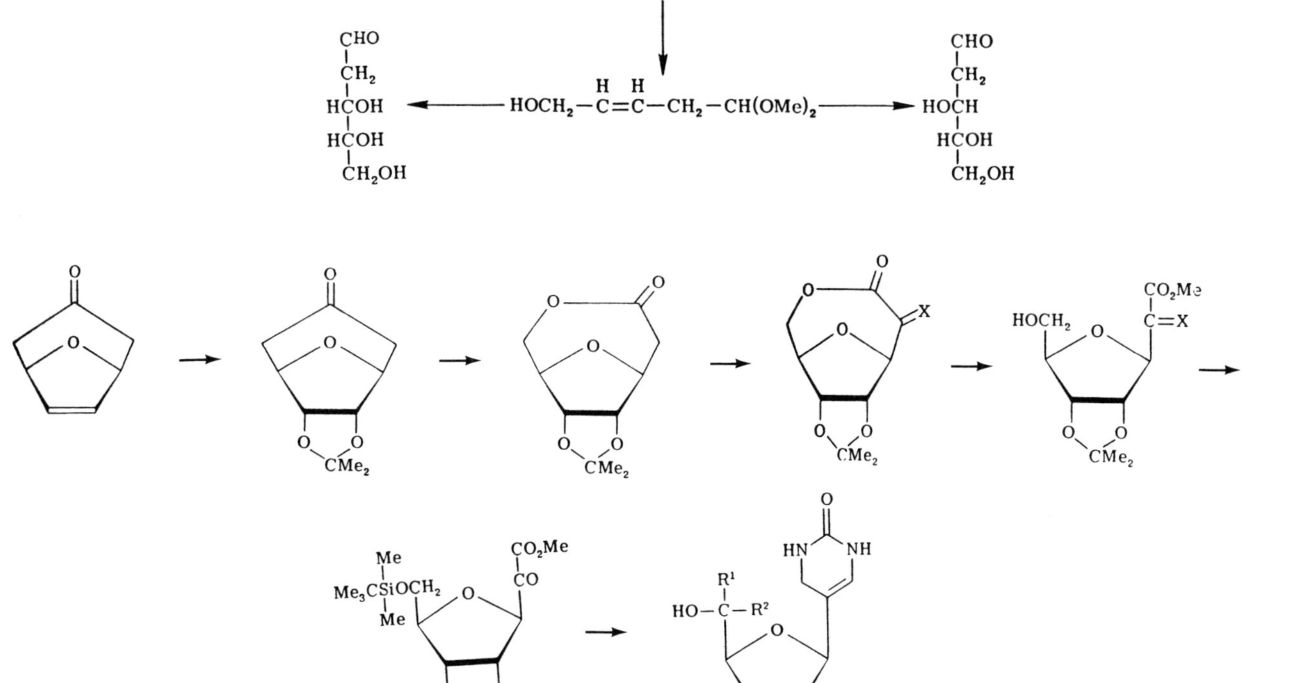

[R. Noyori, *Acc. Chem. Res.* **12**, 61 (1979).]

diol. The resulting *cis*-enol ether was converted to the diethyl acetal, hydroxylated with potassium permanganate, and hydrolyzed to the desired DL-apiose.

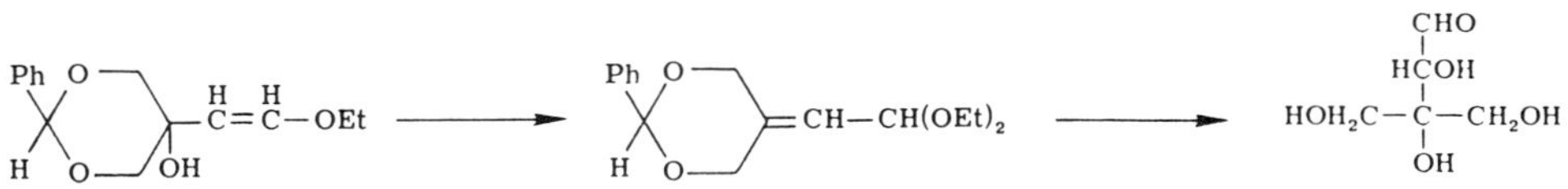

Ph O H O OH H H C=C—OEt → Ph O H O =CH—CH(OEt)$_2$ → CHO HCOH HOH$_2$C—C—CH$_2$OH OH

Racemic mycarose (2,6-dideoxy-3-*C*-methyl-*ribo*-hexopyranose) was synthesized by R. Woodward from the dimethyl acetal of 3-hydroxy-3-methyl-4-hexynal. Partial hydrogenation in the presence of palladium–charcoal afforded a *cis*-alkene, which was hydroxylated with potassium permanganate to give two stereoisomeric triols. Glycosidation with methanolic hydrogen chloride gave a mixture of methyl glycosides that included the desired methyl DL-mycaroside (methyl 2,6-dideoxy-3-*C*-methyl-*ribo*-hexopyranoside).

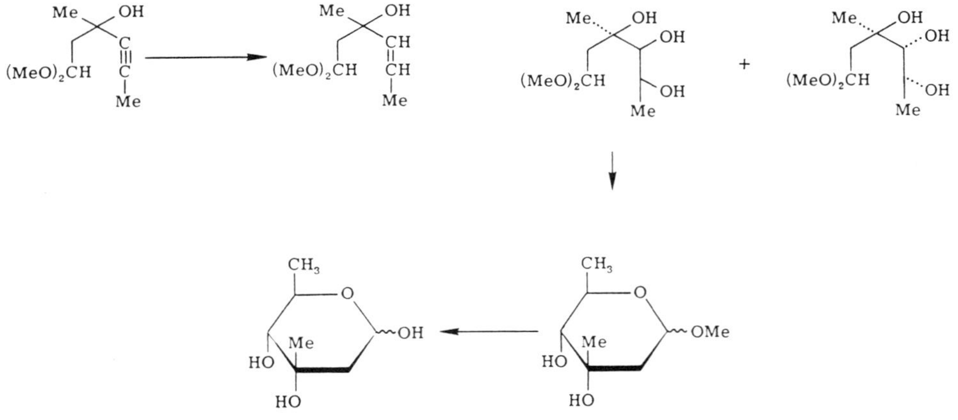

Me OH (MeO)$_2$CH C≡C Me → Me OH (MeO)$_2$CH CH=CH Me → Me OH OH (MeO)$_2$CH OH Me + Me OH OH (MeO)$_2$CH OH Me

↓

CH$_3$ O OH Me HO HO ← CH$_3$ O OMe Me HO HO

A synthesis of branched-chain monosaccharides is based on the finding that 3-furoic acid readily undergoes Birch reduction to give 2,3-dihydro-3-furoic acid. Treatment of the methyl ester with methanol and acid gives, in quantitative yield, methyl tetrahydro-5-methoxy-3-furoate. Bromination of the mixture of isomers, followed by dehydrobromination, gives a 2,5-dihydrofuran derivative, which, by successive hydroxylation, isopropylidenation, and reduction, gives the desired methyl 3-*C*-(hydroxymethyl)-2,3-*O*-isopropylidene-β-DL-erythrofuranoside.

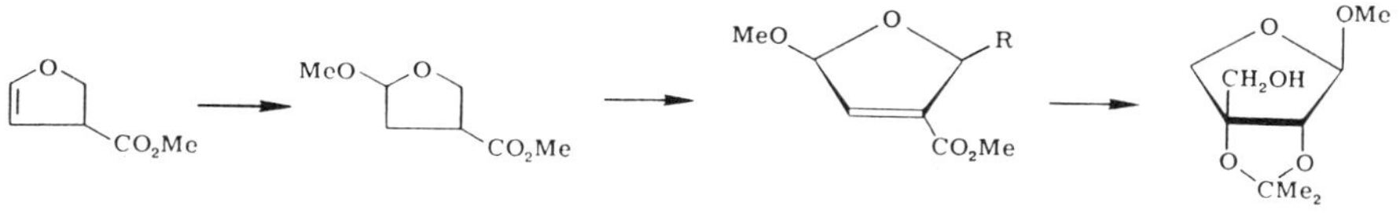

d. *Synthesis of Hexoses and Deoxyhexoses*

(i) FROM UNSATURATED ALDEHYDES AND UNSATURATED ACIDS. The amino sugar daunosamine (3-amino-2,3,6-trideoxy-L-*lyxo*-hexose), which is found in the antibiotic daunorubicin, was prepared in the racemic form from non-sugar precursors (compare this synthesis with Horton's synthesis described on p. 177). The acetal of *trans*-4-hexenal was subjected to allylic amination, followed by treatment with isopropyl alcohol in the presence of hydrochloric acid. Acetylation afforded two isomeric glycosides, from which the DL form of daunosamine was isolated in 70% yield.

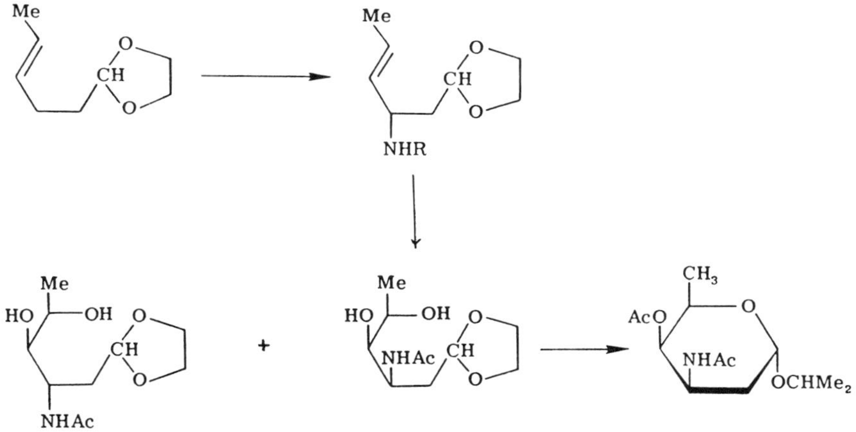

[I. Dyong and R. Wiemann, *Angew. Chem.* **90**, 728 (1978).]

Esters of *cis*-2,4-hexadienoic acid and *cis*-2,5-hexadienoic acid (sorbic and parasorbic acids) have been used to prepare other deoxyhexoses. For example, an analog of daunosamine, 3-acetamido-3,6-dideoxy-*arabino*-hexopyranose, was synthesized from methyl sorbate by treatment with peroxyacetic acid to form an epoxide ring at the site of the 2,3-double bond. Treatment with ammonia opened the ring, introduced the 3-amino function, and formed an amide group. The latter was hydrolyzed and reduced with diisobutylaluminum hydride (DIBAH) to give the desired product. If the epoxide is hydrolyzed and the resulting *trans*-3,6-dideoxy-

DL-*erythro*-2-hexenoic acid is hydrogenated and partially reduced with diisobutyaluminum hydride, the furanose and pyranose forms of DL-amicetose (2,3,6-trideoxy-DL-*erythro*-hexose) is obtained. Optically pure daunosamine was also prepared by Hanessian and Kloss using noncarbohydrate precursors.

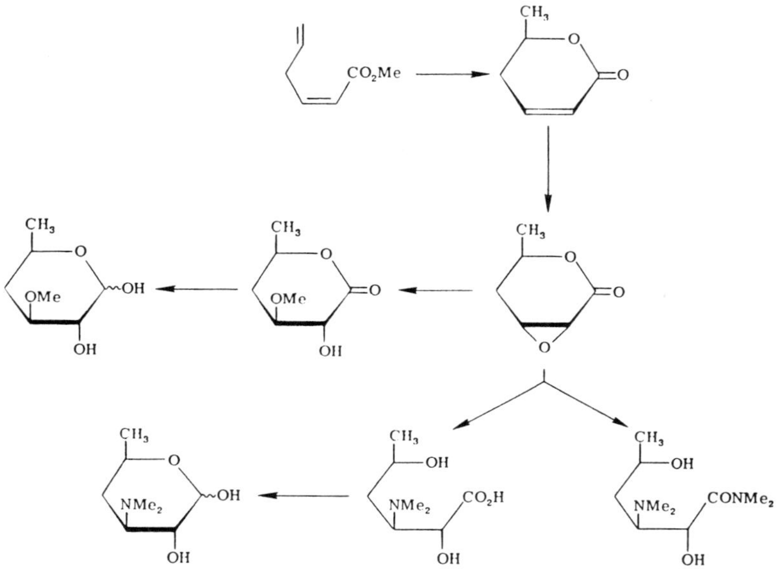

(ii) FROM DIHYDRO-2*H*-PYRANS. It is advantageous to prepare pyranoses from 3,4-dihydro-2*H*-pyran and 5,6-dihydro-2*H*-pyran rather than 2*H*- and 4*H*-pyrans, which are unstable. For example, 2-ethoxy-3,4-dihydro-6-methyl-2*H*-pyran can undergo hydroboration of the double bond, followed by oxidation to ethyl 2,3,6-trideoxy-DL-*erythro*-hexopyranoside. 2-Alkoxy-5,6-dihydro-2*H*-pyrans are also useful derivatives for the synthesis of deoxypentoses and deoxyhexoses. They are synthesized by reaction of 3,4-dihydro-2*H*-pyran with bromine, followed by elimination of the elements of hydrogen bromide with bases. Alternatively, they may be prepared by Diels–Alder reactions—for example, the cycloaddition of 1-alkoxy-1,3-butadienes with such dienophiles as formaldehyde or esters of glyoxylic (oxoethanoic) or mesoxalic (oxopropanedioic) acids. The alkoxyl group in these compounds, being part of the acetal and allylic system, can readily be displaced by other groups in acid-catalyzed or thermal reactions with appropriate alcohols or amines. Further, the double bond will exhibit its normal properties and electrophilic reagents (hydrogen chloride, bromine, etc.) can readily be added. Also, nucleophilic reagents, alcohols, and acetic acid may be added across the double bond under acid catalysis.

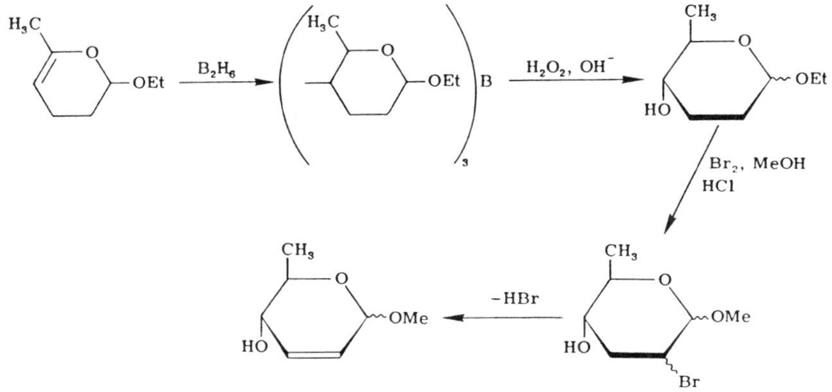

[Y. Yasuda and T. Matsumoto, *Tetrahedron Lett.* pp. 4393, 4397 (1969).]

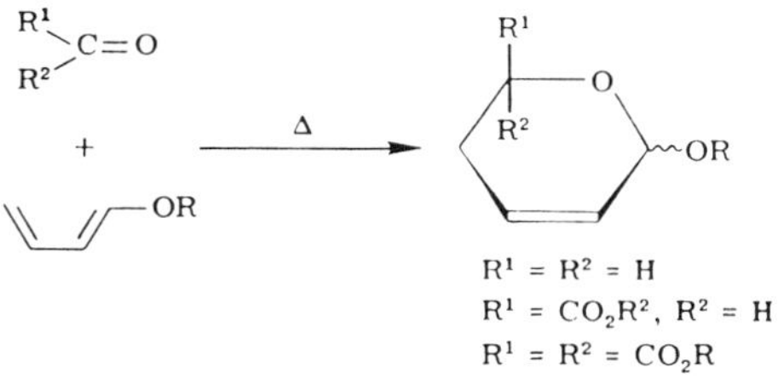

$R^1 = R^2 = H$
$R^1 = CO_2R^2,\ R^2 = H$
$R^1 = R^2 = CO_2R$

(iii) FROM BICYCLIC COMPOUNDS. Diels–Alder adducts of furan and methyl 2-nitroacrylate can be hydroxylated to give a pair of exo-*cis*-diols. The isopropylidene derivative of the first is treated with diazabicyclo[5.4.0]undec-5-ene (DBU) to give two alkenes, followed by ozonolysis and reduction to give a mixture of the epimeric triols. These are cleaved with periodate to yield 2,5-anhydro-3,5-*O*-isopropylidene-DL-allose.

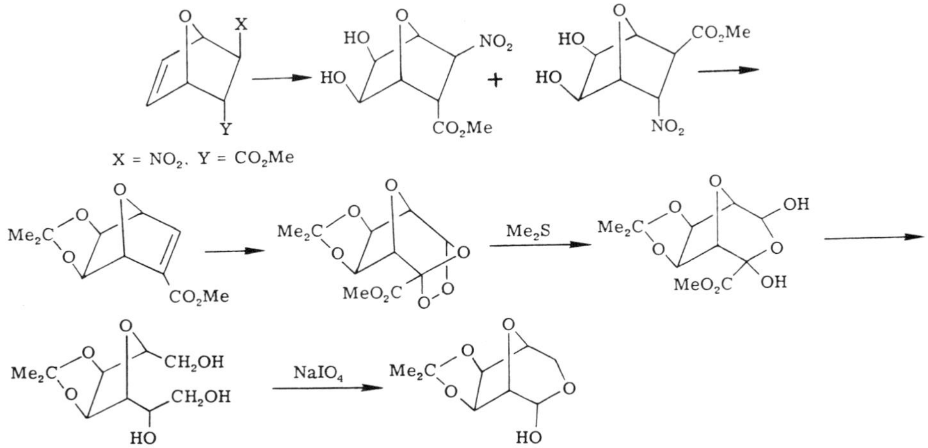

$X = NO_2$, $Y = CO_2Me$

Me_2S

$NaIO_4$

[G. Just and A. Martel, *Tetrahedron Lett.* p.1517 (1973).]

PROBLEMS

1. Outline the synthetic sequence of reactions that would lead to the following labeled compounds:
   (a) D-Glucose labeled with ^{14}C on C-1.
   (b) D-Glucose labeled with ^{18}O attached to C-1.
   (c) L-Ascorbic acid, having one ^{2}H attached to C-6 (CHDOH).
   (d) 3-Amino-3-deoxy-D-glucose labeled with ^{15}N.
2. Explain the following:
   (a) Why methyl glycosides readily hydrolyze in acid media and are stable in alkaline media.
   (b) Why methyl 2-deoxy-D-*arabino*-hexopyranoside is hydrolyzed faster than methyl D-glucopyranoside.
   (c) Why methyl α-D-ribofuranoside yields a 2,3-*O*-isopropylidene derivative instead of a 4,6 derivative.
3. Show how the following conversions may be carried out (more than one reaction may be involved in each step).

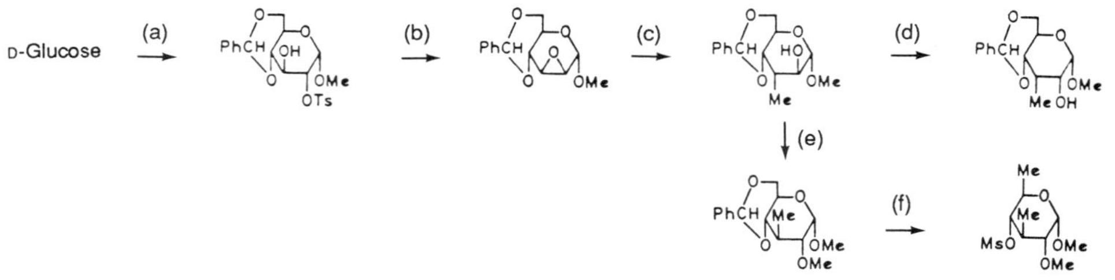

D-Glucose (a) → (b) → (c) → (d) →

(e)

(f) →

II

Oligomeric Saccharides: Oligosaccharides and Nucleotides

6

Structure of Oligosaccharides

Oligosaccharides are polymeric saccharides that have, as their name denotes, a low degree of polymerization (*oligos* means "few" in Greek). They are composed of 2–10 glycosidically linked monosaccharides, which can be liberated by depolymerization (for example, by acid hydrolysis). Oligosaccharides having degrees of polymerization of 2–3 are sweet in taste and are included among sugars, whereas higher members are devoid of taste and are not referred to as sugars.

Oligosaccharides are grouped into simple (or true) oligosaccharides, which on depolymerization yield monosaccharides *only*, and conjugate oligosaccharides, which are linked to such nonsaccharides as peptides and lipids and on depolymerization yield monosaccharides and aglycons. The simple oligosaccharides are further classified according to (a) degree of polymerization, into disaccharides, trisaccharides, tetrasaccharides, etc.; (b) whether they are composed of one or more types of monosaccharides, into homo- and hetero-oligosaccharides; and finally (c) whether they do or do not possess a hemiacetal function at one terminus of the molecule, into reducing and non-reducing oligosaccharides.

Related homo-oligosaccharides can form homologous series; a homologous series of oligosaccharides is a group of similarly linked oligosaccharides that are composed of the same monomer, and whose degree of polymerization increases in the series by one unit at a time. When homopolysaccharides are partially hydrolyzed, they often yield homologous series of oligosaccharides. For example, the malto-oligosaccharides obtained by partial hydrolysis of starch comprise dimers, trimers, tetramers, etc., composed of α-D-glucopyranosyl units linked by $(1 \rightarrow 4)$ glycosidic bonds.

The complete structure of oligosaccharides is established when the following points are determined: (a) the degree of polymerization, i.e., the number of monosaccharide units present in the oligomer molecule; (b) the nature of the monosaccharide monomer(s); (c) in the case of hetero-oligosaccharides, the monosaccharide sequence; (d) the ring size (pyranose or furanose) and the position of linkage of the different monosaccharides $(1 \rightarrow ?)$; and (e) the anomeric configuration (α or β) and the conformation of the monosaccharide units.

I. DETERMINATION OF THE DEGREE OF POLYMERIZATION

The following procedures are recommended for determining the degree of polymerization (DP) of the oligosaccharides.

A. Mass Spectrometry

If the molecular weight of the oligosaccharide is less than 1000, mass spectrometry may be used to determine molecular weight and hence degree of polymerization. It must be remembered, however, that mass spectra produced by electron impact seldom show the molecular ions of saccharides and that other ionization methods, such as fast atom bombardment (FAB) or chemical ionization, must be used to reveal these ions. The pertrimethylsilyl ethers of oligosaccharides or of their alditols (obtained by reduction and silylation) are often used for mass spectrometric measurements, because they are more volatile than free oligosaccharides. Alternatively, peracetylated disaccharides, or other esters of disaccharides or their glycosides, may be used in the chemical ionization mode. Figure 1 shows three mass spectra of a disaccharide glycoside obtained by (a) negative chemical ionization, (b) electron impact, and (c) positive chemical ionization (CI). Only the CI mass spectra (negative and positive) revealed molecular ions.

B. Chromatography

It is often possible to determine the degree of polymerization of members of a homologous series of polymeric saccharides from their position

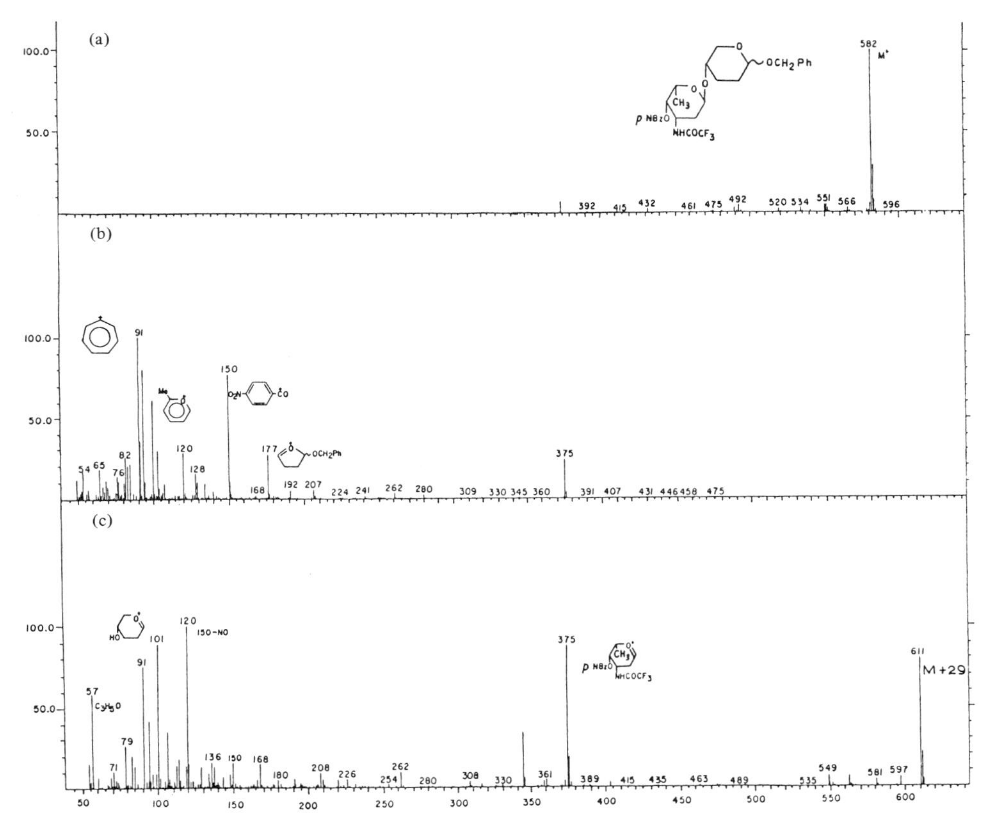

Fig. 1. Mass spectra of a disaccharide obtained by (a) negative-ion chemical ionization in methane, (b) electron impact, and (c) positive-ion chemical ionization in methane.

on chromatograms relative to that of a known member of the series. For example, partial hydrolysis of starch yields D-glucose and a homologous series of malto-oligosaccharides consisting of di-, tri-, tetra-, and penta-saccharides, etc. This mixture is separated in chromatographs according to degree of polmerization. On paper chromatograms and liquid chromatograms, the mobility of the saccharides is inversely proportional to their DP (i.e., the largest oligomers move slowest), and on gel filtration chromatograms it is directly proportional to DP (i.e., the largest oligomers move fastest). Of considerable importance in the separation of large oligosaccharides and polysaccharides are the Sephadex and Sepharose products, which are used in gel filtration and ion exchange. The fact that the oligomers are eluted in the order of increasing or decreasing DP makes it possible to determine their DP by determining their order of elution relative to that of a known member. Thus, by recognizing maltose (and glucose) in chromatograms of partially hydrolyzed starch (see Fig. 2), it is possible to identify the subsequent bands as maltotriose, maltotetraose, maltopentaose, etc.

C. Reducing Power

The degree of polymerization of reducing oligosaccharides, particularly those having a low molecular weight, can be determined from their reducing power. For example, the reducing power of a malto-oligosaccharide, relative to glucose (taken as 100%), is matched with the calculated values for oligomers having different degrees of polymerization. Thus, a disaccharide would be expected to have about half the reducing power of glucose (actually, 53%), a trisaccharide a third (actually 35%), a tetrasaccharide a fourth (26%) that of glucose, etc. Accordingly, if a malto-oligosaccharide exhibits a reducing power equal to 35.8% of that of D-glucose, it can be safely assumed that it is a trisaccharide. The differences in the reducing powers of successive members of a homologous series of oligosaccharides decrease with increasing DP, and beyond a DP of 5 the differences become too small for reliable DP measurements by this method.

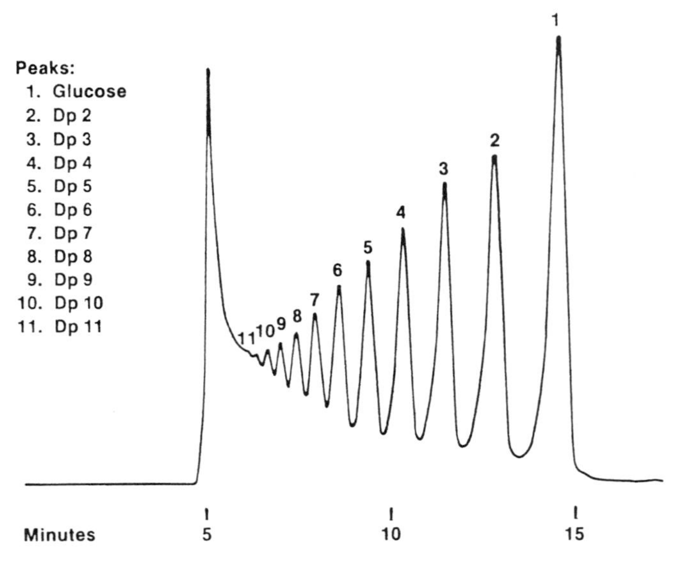

Fig. 2. Liquid chromatogram of partially hydrolyzed starch, showing glucose (1), maltose (2), and the higher malto-oligosaccharides (3–10).

II. MONOSACCHARIDE COMPONENTS

The nature of the monomers in an oligosaccharide is established by identifying the monosaccharides liberated by acid-catalyzed hydrolysis. Homo-oligosaccharides, which are composed of one type of monosaccharide, yield only one monosaccharide, which can be isolated from the hydrolysate by conversion into crystalline derivatives. On the other hand, hetero-oligosaccharides afford, on hydrolysis, a mixture of monosaccharides, which must be separated by chromatography. Paper chromatography, thin-layer chromatography (TLC), or liquid chromatography (LC) can be used without pretreating the monosaccharides, but a gas chromatographic (GC) separation requires prior pertrimethylsilylation of the saccharides to give ethers that are volatile. It should be noted that monosaccharides that have been silylated appear as double peaks (one peak for the α anomer and one for the β), which tends to crowd the chromatograms. To avoid this complication, the monosaccharides may be reduced before silylation (silylated alditols appear as single peaks). It should also be noted that most available columns cannot differentiate between D and L enantiomers. This will change with the availability of commercial columns filled with chiral supports, which will render the separation of enantiomers possible. Until then, polarimetric measurements are needed [optical rotation, optical rotatory dispersion (ORD), or circular dichroism (CD)] to determine the configuration of the monosaccharide component(s) of oligosaccharides.

III. MONOSACCHARIDE SEQUENCE

Homo-oligosaccharides (whether reducing or nonreducing) do not require a monosaccharide sequence determination, because they have identical monomers all through their chains. Hetero-oligosaccharides, except nonreducing disaccharides, must have the sequence of their monomers determined. In the case of reducing heterodisaccharides, the monomer at the nonreducing end of the molecule differs from that at the reducing end, and the position of both monomers must be determined. It was stated above that nonreducing disaccharides do not require a monosaccharide sequence determination. This is because the two monomers are situated at two similar (nonreducing) ends of a chain that has no beginning (no reducing terminus). On the other hand, a nonreducing hetero-oligosaccharide with DP $>$ 2 is made up of a monosaccharide linked glycosidically to the hemiacetal function of a reducing oligosaccharide whose sequence must be determined.

The methods available for determining the monomer sequence are presented here in order of increasing DP of the oligosaccharides, starting with disaccharides.

A. Monosaccharide Sequence in Disaccharides

The sequence of monomers in disaccharides is determined when the monomer situated at the reducing terminus of the molecule is identified. This automatically determines the position of the other monomer (it must be at the nonreducing end). To identify the monosaccharide at the reducing terminus, use is made of the fact that this saccharide moiety exists in equilibrium with an acyclic form and is, therefore, much more susceptible

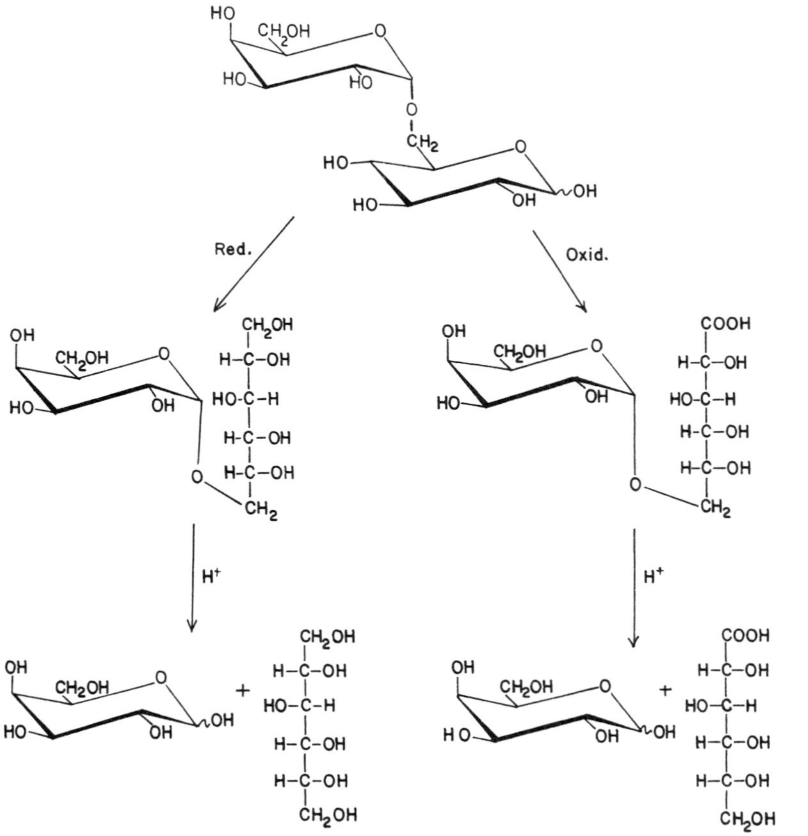

Red. Oxid.

H^+ H^+

D-Galactose D-Glucitol D-Galactose D-Gluconic acid

to oxidants and reducing agents than the other moiety. Thus, mild oxidation converts reducing disaccharides into aldobionic acids, which on hydrolysis give an aldonic acid (from the reducing terminus) and a reducing monosaccharide (from the nonreducing end). Reduction gives an aldobiitol, which on hydrolysis yields an alditol from the reducing moiety and a monosaccharide from the nonreducing moiety. The use of this method of structure determination is exemplified by the oxidation and reduction of melibiose (6-*O*-α-D-galactopyranosyl-D-glucopyranose) to give, in the first case, melibionic acid (6-*O*-α-D-galactopyranosyl-D-gluconic acid) and, in the second, melibiitol (6-*O*-α-D-galactopyranosyl-D-glucitol). Hydrolysis of these yields D-galactose, from the nonreducing end of the molecule, and D-gluconic acid in the first case and D-glucitol in the second (both formed from the reducing half of the molecule). This experiment shows that D-galactose is the monomer located at the nonreducing end of the dimer and that D-glucose is the one at the reducing end.

B. Monosaccharide Sequence in Trisaccharides

Because it is easier to determine the structure of disaccharides than that of trisaccharides, it is often advantageous to deduce the structure of the latter by identifying (or determining the structure of) the two disaccharides resulting from partial hydrolysis of the trisaccharide under investigation. If known, the disaccharides would be compared with authentic samples, and if not, the new disaccharides would be investigated as just shown. In piecing together the structure of a trisaccharide from those of two disaccharides, the following rules may be used.

(i) If the trimer is composed of three (different) types of monomer (for example, A–B–C), then only one monomer, namely B, will be found in both dimers (A–B and B–C). This monomer must be located in the center of the trimer, and the remaining monomers (A and C) occupy the same terminal positions in the trimer as they do in the dimer [the monomer present at the nonreducing end of a disaccharide (A) will be at the nonreducing end of the trisaccharide, and the monomer found at the reducing end of a disaccharide (C) will be at the reducing end of the trisaccharide].

(ii) If the trimer is composed of only two types of monomer (for example A–A–B, A–B–B, A–B–A, or B–A–B), it is possible to have one or two common monomers in the resulting dimers. If only one monomer is common to the two dimers, for example, A in A–A and A–B (obtained from A–A–B) or B in A–B and B–B (obtained from A–B–B), the same rules are used. However, if both monomers (A and B) are common to the two dimers (for example, A–B and B–A, obtained from either A–B–A or B–A–B), an estimation (by GC or LC) of the two monomers in the trisaccharide hydrolysate will reveal that one monomer (A in A–B–A and B in B–A–B) occurs in twice the amount of the other; this monomer must be terminally located in the trisaccharide molecule. To illustrate this method, a nonreducing trisaccharide, raffinose, will be used as an example of a trisaccharide having three different monomers (A–B–C). This trisac-

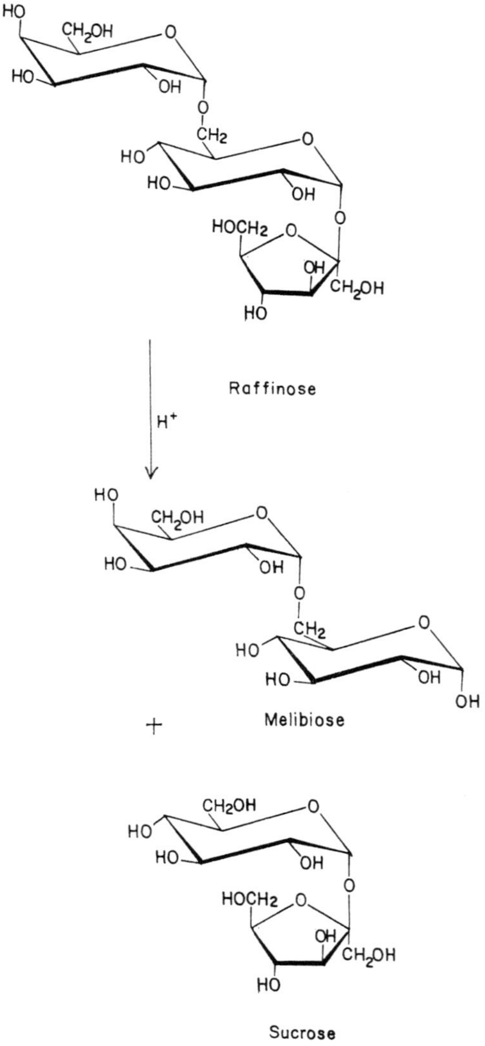

Raffinose

H^+

Melibiose

+

Sucrose

charide yields on partial hydrolysis a reducing disaccharide (6-*O*-α-D-galactopyranosyl-D-glucopyranose) and the nonreducing disaccharide sucrose (α-D-glucopyranosyl β-D-fructofuranoside). The common monomer in both disaccharides is D-glucopyranose, which must be at the middle of the trimer. The other two monomers (D-galactopyranose and D-fructofuranose) are then assigned positions at the two ends of the nonreducing trimer. Oxidation (or reduction) of the first disaccharide, followed by hydrolysis, reveals that D-glucose is at the reducing end and D-galactose at the nonreducing end of the dimer molecule and confirms the location of the latter at the nonreducing end of the trimer molecule. Because the second disaccharide is nonreducing, the terminal D-fructofuranose must be attached through O-2 to O-1 of D-glucose. This establishes the structure of raffinose as O-α-D-galactopyranosyl-(1 → 6)-α-D-glucopyranosyl β-D-fructofuranoside.

C. Monosaccharide Sequence in Tetrasaccharides

The sequence in tetrasaccharides can be deduced by determining the sequence of monosaccharides in the oligosaccharides that result from their partial hydrolysis (two trisaccharides and three disaccharides). If the tetrasaccharide is composed of four different monomers (A–B–C–D), the monosaccharide sequence may be deduced by determining the monomer sequence in the disaccharides (A–B, B–C, and C–D) produced by partial hydrolysis. Use is made of the fact that the monomers common to two dimers (B, found in A–B and B–C; and C, in B–C and C–D) must be linked together and must be located in the center of the tetramer (thus, B must be attached to A and C, and C to B and D). The same rule applies if only one monomer is repeated and the two analogous monomers are contiguous (as determined from the fact that *one* of the dimers is a homodisaccharide). Examples of such homodisaccharides are A–A, obtained from A–A–B–C; B–B, obtained from A–B–B–C; and C–C, produced from A–B–C–C. If, on the other hand, the repeating monomers are not contiguous but are separated by one monomer (as in A–B–A–C and A–B–C–B) or by two monomers (as in A–B–C–A), a study of the dimers alone may not suffice to elucidate the structure of the tetramer, making it necessary also to study the structure of the two trisaccharides resulting from it by partial hydrolysis (see above).

In general, the structure elucidation of polymers becomes more difficult as the degree of polymerization increases. However, because higher oligosaccharides are composed of repeating units seldom larger than tetrasaccharide fragments (and often composed of mono-, di-, or trisaccharides), the number of possible oligosaccharides liberated on partial hydrolysis remains relatively small. Thus, whereas a tetrasaccharide (A–B–C–D) will give on partial hydrolysis two trisaccharides (A–B–C and B–C–D) and three disaccharides (A–B, B–C, and C–D), an octasaccharide (or, for that matter, a polysaccharide) composed of the same tetrasaccharide repeating unit will give only two additional trisaccharides (D–A–B and C–D–A) and one additional disaccharide (D–A).

IV. RING SIZE AND POSITION OF LINKAGE

The next stage in the structure elucidation of oligosaccharides involves determination of the ring size and the position of linkage of the monosaccharide constituents. For nonreducing disaccharides, the position of linkage must be between anomeric carbon atoms, as otherwise the disaccharide would be reducing. If the nonreducing disaccharide is composed of two aldoses, these two monomers must be linked by an acetal oxygen bridge joining C-1 of one aldose to C-1 of the other, which is signified by $(1 \leftrightarrow 1)$; if the disaccharide is composed of two 2-glyculoses (2-ketoses), the oxygen bridge must link C-2 of one 2-glyculose to C-2 of the other $(2 \leftrightarrow 2)$; finally, if the disaccharide is composed of one aldose and one 2-glyculose, the linkage is $1 \rightarrow 2$ (with the oxygen bridge linking C-1 of the aldose to C-2 of the 2-glyculose). Accordingly, when the monosaccharide components of a nonreducing disaccharide are identified, there is no uncertainty about the positions of linkage, but only about the size of the rings (and the anomeric configuration, which will be discussed later). Reducing disaccharides, on the other hand, must have their position of linkage and ring size determined. The position of linkage or ring size, or both, of the monosaccharide components of reducing and nonreducing oligosaccharides may be determined by labeling or by partial hydrolysis, as follows.

A. Labeling of Free Hydroxyl Groups

The free hydroxyl groups in oligosaccharides are attached to carbon atoms that are not involved in ring formation or in glycosidic bonds. These carbon atoms can be recognized by marking them with a suitable label, for example, by attaching to their oxygen atoms permanent blocking groups that would resist the conditions (heat and acid) needed for hydrolysis of the labeled oligosaccharide. Methylation is often used, because many of the partially methylated monosaccharides that result from such hydrolysis have been characterized by GC. Methylation and hydrolysis of a reducing disaccharide yields two methylated monosaccharides; the first has two free hydroxyl groups (at C-1 and where the ring was attached) and the second has three free hydroxyl groups (at C-1, where the ring was attached, and where the glycosidic bond was attached). The vacant positions in the latter might create ambiguity, as it is not known which positions involve the ring and which the glycosidic bond. To avoid this confusion, it will be necessary to carry out another methylation on an acyclic disaccharide derivative—for example, the aldobionic acid ob-

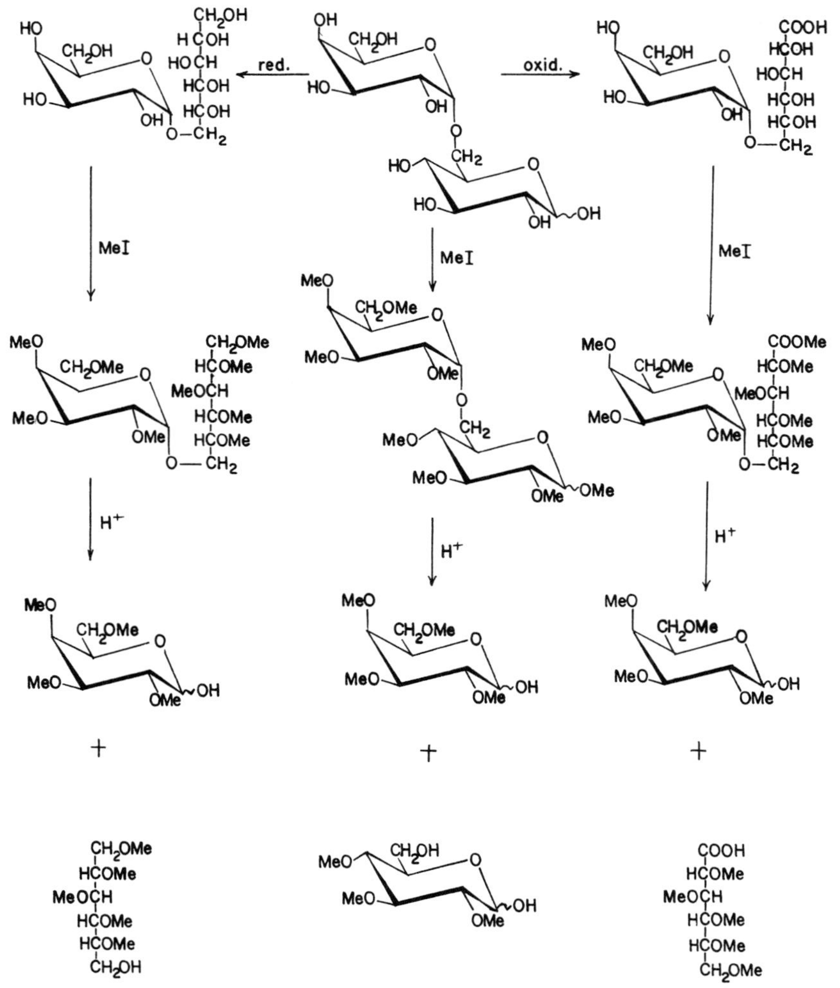

tained by oxidation of the disaccharide, or the aldobiitol obtained by reduction. These compounds have one acyclic moiety (the aldonic acid and the alditol, respectively), so that, after hydrolysis of the alkylated dimer, the unmethylated position in the acyclic moiety will be the position of the glycosidic linkage.

If the disaccharide is nonreducing, the position of the glycosidic linkage is known (C-1 for an aldose and C-2 for a 2-glyculose), and only one methylation experiment is needed (the unlabeled position is where the ring was attached). The use of labeling in structure elucidation is exemplified by (1) the methylation and hydrolysis of melibiose to yield 2,3,4,6-tetra-*O*-methyl-D-galactose and 2,3,4-tri-*O*-methyl-D-glucose and (2) the oxidation of melibiose to melibionic acid or its reduction to melibiitol, then methylation of these, followed by hydrolysis of the methylated aldobionic acid or aldobiitol to yield 2,3,4,6-tetra-*O*-methyl-D-galactose and 2,3,4,5-tetra-*O*-methyl-D-gluconic acid in the first case, and 1,2,3,4,5-penta-*O*-methyl-D-glucitol in the second.

B. Partial Hydrolysis

It is possible to use the structures of known disaccharides to determine those of higher oligosaccharides. Thus, the fact that the trisaccharide raffinose (already discussed) yields on partial hydrolysis melibiose and sucrose, whose structures are known to be 6-*O*-α-D-galactopyranosyl-D-glucopyranose and α-D-glucopyranosyl β-D-fructofuranoside, respectively, establishes that the trisaccharide molecule is composed of an α-D-galactopyranose ring attached through an α-glycosidic bond to O-6 of an α-D-glucopyranose ring, which is in turn attached glycosidically to the anomeric position of a β-D-fructofuranose ring. In other words, the trisaccharide must be 6-*O*-α-D-galactopyranosyl-α-D-glucopyranosyl β-D-fructofuranoside.

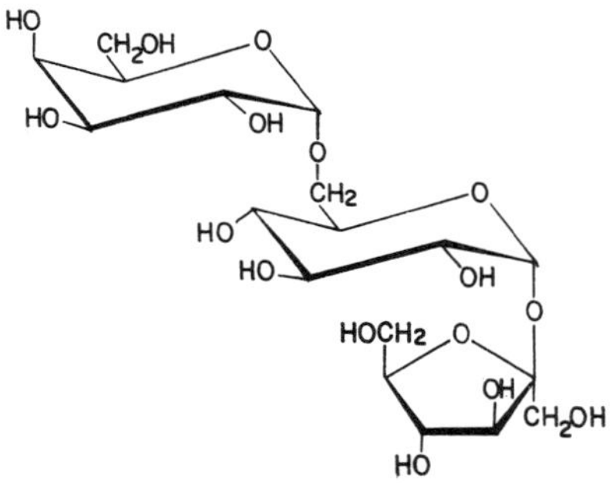

V. ANOMERIC CONFIGURATION AND CONFORMATION OF THE SACCHARIDE

A. By Partial Hydrolysis

It was mentioned earlier that on partial hydrolysis raffinose yields melibiose and sucrose, which are 6-*O*-α-D-galactopyranosyl-D-glucopyranose and α-D-glucopyranosyl β-D-fructofuranoside, respectively. This suggests that the linkage between the galactose and the glucose units in the trisaccharide is α-D, and that between glucose and fructose is α-D for glucose and β-D for fructose.

B. By Nuclear Magnetic Resonance Spectroscopy

The method most commonly used for determining both the anomeric configuration and the conformation of oligosaccharide units is NMR spectroscopy. The coupling constant of the anomeric proton ($J_{1,2}$) is measured and used to determine the dihedral angle between H-1 and H-2 in the disaccharide. Because this angle depends not only on the anomeric configuration but also on the conformation, the two are determined concurrently. The sample is irradiated at the frequency of H-1, so that the signal of H-2 is identified (as it is partly decoupled), and $J_{2,3}$ is determined. The procedure is repeated by irradiating the sample at the frequency of the H-2 absorption (to identify the signal due to H-3 and to measure $J_{3,4}$) and then at the frequency of H-3, to identify the signal due to H-4, etc. When all of the coupling constants of one ring have been measured, the process is repeated for the other ring. Finally, the Karplus equation is used to estimate the dihedral angle between the different protons, which establishes the conformation of the unit and its anomeric configuration. It should be noted that, in order to identify all of the signals in an oligosaccharide spectrum, high-resolution NMR instruments are needed (preferably ones having two-dimensional mapping capabilities). In the absence of such equipment, it is still possible to determine the anomeric configuration and the conformation by measuring the coupling constants of H-1 and H-4 (for pyranose rings). Figure 3 shows the NMR spectrum of octa-*O*-acetyl-α-D-glucopyranosyl α-D-glucopyranoside (α,α-trehalose octaacetate), which clearly shows that this molecule possesses two identical α-D-glucopyranosyl units in the $^4C_1$(D) conformation. This is apparent from the coupling of the anomeric proton and the fact that the two rings produce identical signals, as well as from the coupling of H-4 (split by the two trans-diaxial protons, at H-3 and H-5).

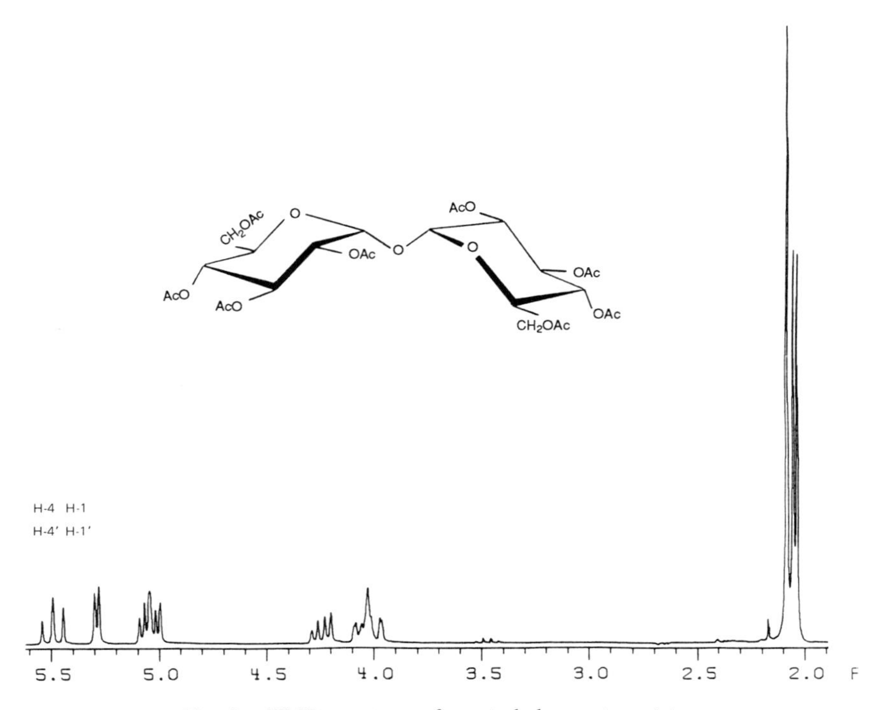

Fig. 3. NMR spectrum of α,α-trehalose octaacetate.

Higher oligosaccharides usually require two-dimensional NMR measurements to analyze their complicated proton spectra. Homonuclear or heteronuclear correlations may be used to achieve this result. Figure 4 shows a $^{1}H-^{13}C$ heteronuclear correlation for a branched tetrasaccharide composed of a D-glucopyranosyl unit attached to a chain composed of three D-xylopyranose units. The protons are designated by their position numerals followed, in certain cases, by primes to designate the monosaccharide to which they are attached. For example, 1 designates the anomeric proton of the xylose unit located at the reducing end of the molecule, 1′ that of the next xylosyl unit, 1″ that of the last xylosyl unit, and 1‴ the anomeric proton of the glucopyranosyl unit.

C. By Crystallography

Another way to determine the anomeric configuration and the conformation is by crystallography (using either X-ray or neutron diffraction). Both techniques give the complete structure (in the solid state) of a crys-

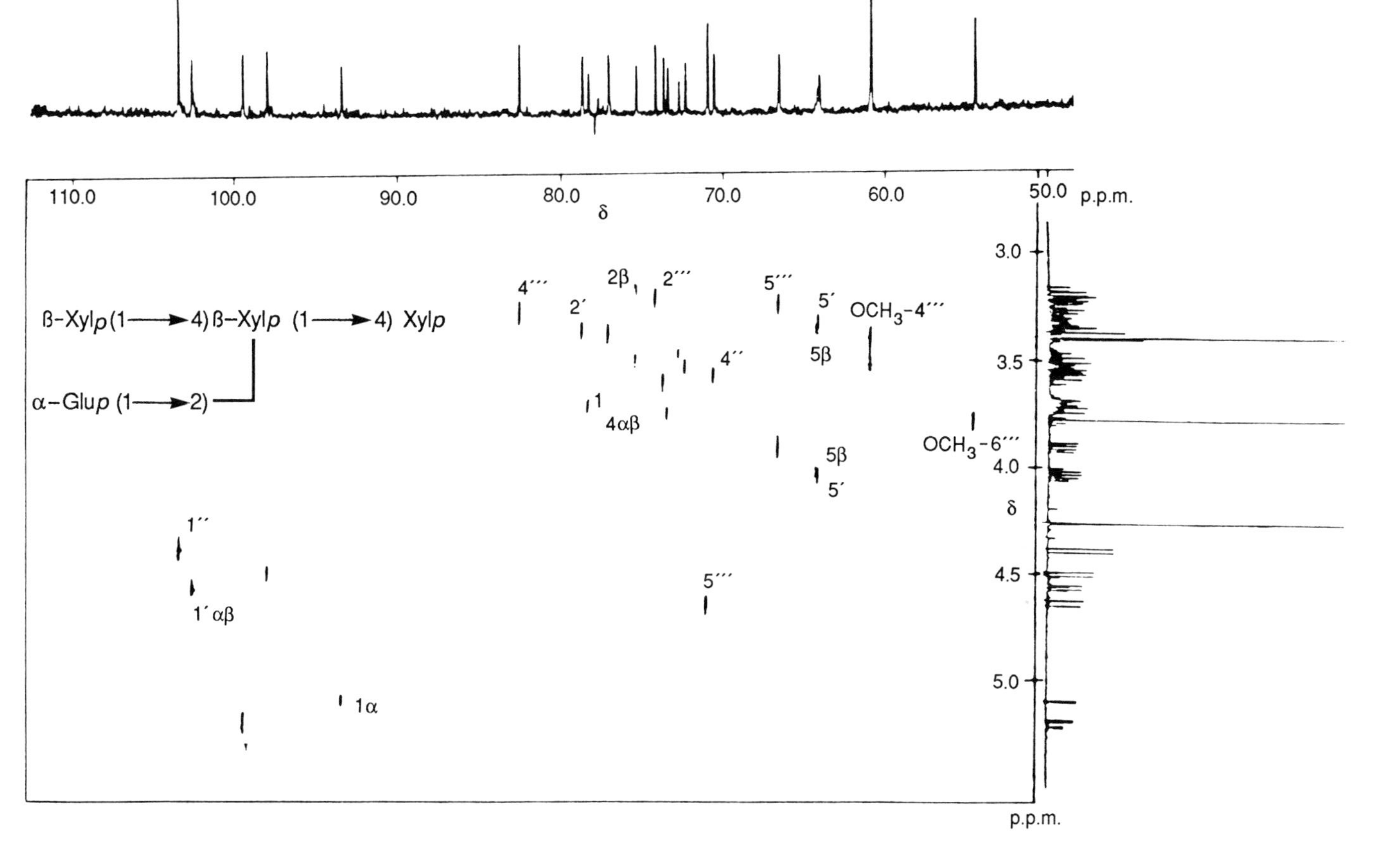

Fig. 4. Two-dimensional NMR of a tetrasaccharide.

talline oligosaccharide, including the orientation of the two rings vis-à-vis one another (angles ϕ and ψ).

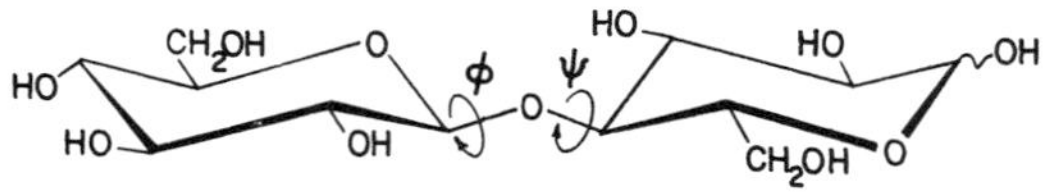

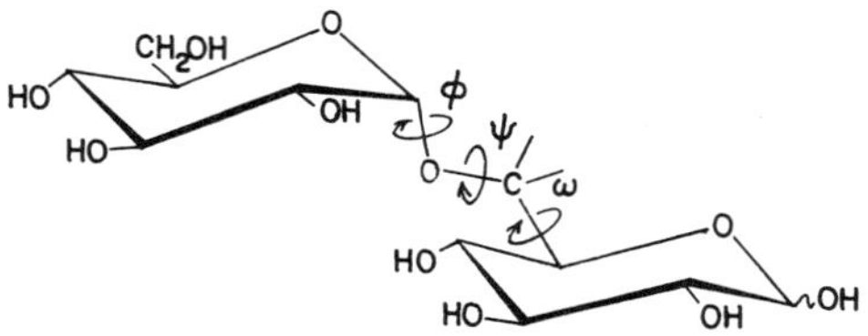

Figure 5 is a diagram, deduced from X-ray diffraction, of a dihydrate of α,α-trehalose. The anomeric configuration and the ${}^4C_1$ conformation of the two rings are clearly revealed. The remarkable symmetry of the molecule, which agrees with the NMR data discussed earlier, may also be noted.

D. Enzymatic Method

The orientation of the glycosidic bond of disaccharides can be determined by enzymatic hydrolysis. For example, emulsin, an enzyme obtained from bitter almonds, is known to hydrolyze β-D-glucosidic linkages and not α-D linkages. Accordingly, if a D-glucose containing disaccharide is hydrolyzed by emulsin, it may be concluded that it possesses a β-D linkage.

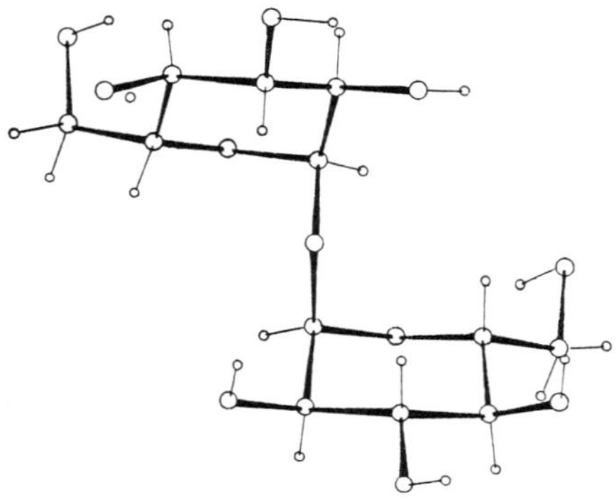

Fig. 5. Conformation of α,α-trehalose dihydrate, determined by X-ray crystallography.

VI. INSTRUMENTAL METHODS FOR STRUCTURE ELUCIDATION

Certain naturally occurring oligosaccharides of biological importance, such as glycoproteins, are isolated in milligram quantities, which are not sufficient for structural investigations by wet chemical methods. Such quantities are, however, adequate for structure elucidation by nondestructive instrumental methods such as single-crystal X-ray (or neutron diffraction) crystallography, which was discussed previously, or high-resolution ^{1}H-NMR spectroscopy, which is treated next.

A. High-Resolution ^{1}H-NMR Spectroscopy*

The structure elucidation of naturally occurring oligosaccharides by ^{1}H-NMR spectroscopy was made possible by the advent of instruments having Fourier transform capabilities, which enabled the use of small samples, and of spectrometers with superconducting magnets, which made possible the high resolution needed for this type of work. The task was facilitated by the following factors. (a) Only a dozen or so monosaccharides occur naturally in oligosaccharides, and a smaller number of these is found in any given group or class of oligosaccharides under investigation (for example, *N*-acetylneuraminic acid and four monosaccharide types are found in glycoproteins). (b) The ^{1}H-NMR spectra of oligosaccharides are composites of the spectra of the different monosaccharide components, which facilitates their recognition. (c) It is easy to distinguish between aldoses and ketoses by ^{1}H-NMR spectroscopy (the former possess anomeric protons, whereas the latter do not). (d) The monosaccharide components of an oligosaccharide usually exist in one conformation ($^{4}C_{1}$ for pyranoses and a twist for furanoses).

Nine aldoses are found in naturally occurring oligosaccharides; they comprise three hexoses (glucose, mannose, and galactose), two amino sugars (*N*-acetyl-D-glucosamine and *N*-acetyl-D-galactosamine), two 6-deoxyhexoses (fucose and rhamnose), and two pentoses (arabinose and xylose). The two ketoses (fructose and sorbose) found in oligosaccharides are readily recognized by their lack of anomeric protons. Also lacking anomeric protons is the ketose *N*-acetylneuraminic acid, a nine-carbon acid found in glycoproteins.

To determine the structure of an oligosaccharide by high-resolution ^{1}H-NMR spectroscopy, three distinct areas of the NMR spectrum are examined.

* For a review of this topic, see J. F. G. Vliegenthart, L. Dorland, and H. van Halbeek, *Adv. Carbohydr. Chem. Biochem.* **41,** 209 (1983).

(a) *The high-field area*. This section of the spectrum reveals methyl group protons. For example, a doublet at δ 1.3 is indicative of a 6-deoxyhexose (fucose or rhamnose); a singlet at δ 2.1 suggests an acetyl group, which may be from an acetylated amino sugar (*N*-acetyl-D-glucosamine or *N*-acetyl-D-galactosamine) or *N*-acetylneuraminic acid. The last would show, in addition, two multiplets arising from the protons of the 3-deoxy function.

(b) *The low-field area*. This area of the spectrum reveals the anomeric protons and is used to distinguish between aldoses (which possess anomeric protons) and ketoses including *N*-acetylneuraminic acid (which lack them). Furthermore, the coupling of the anomeric protons of aldoses can be used to determine their anomeric configuration.

(c) *The central part of the spectrum*. Most of the saccharide protons are situated in this area. It is usually quite crowded with signals, but careful examination may reveal characteristic splitting patterns, which can be used, together with the data obtained from the other two regions of the spectrum, to determine the configuration of the different monomers.

Successful structure elucidation of oligosaccharides by high-resolution ^{1}H-NMR spectroscopy requires knowledge of the characteristic traits in the spectra of the monomers. Consider the protons attached to the six-membered ring of the nine aldopyranoses found in oligosaccharides, and ignore, for the time being, the group attached to C-5 (which may be a second hydrogen atom in the case of pentoses, a methyl group in the case of 6-deoxyhexoses, or a hydroxymethyl group in the case of hexoses). It will be seen that these pyranoses exist in three stereochemical families related to three pentoses, xylose, lyxose, and arabinose. Each group is, of course, further divided into two subclasses according to the anomeric configuration (whether it is α or β). The characteristics of the ^{1}H-NMR spectra of the three families of aldopyranoses are given next, together with the structure of the pentoses to which they are related. The latter are depicted with only the ring protons, so that their NMR spectra may be readily visualized. Also presented is *N*-acetylneuraminic acid, an important component of glycoproteins.

1. The Xylose Family

The D-xylose family comprises *N*-acetyl-D-glucosamine, D-glucose, and D-xylose. Its members are characterized by large coupling constants for H-2, H-3, H-4, and H-5, arising from the trans-diaxial orientation of these consecutive protons. This causes the protons on C-3 and C-4 to appear as triplets having large couplings. The members of this group are differentiated by the presence of an *N*-acetyl group in the case of *N*-acetyl-D-glucosamine and by the splitting pattern of H-4 and H-5 in the case of glucose and arabinose. Furthermore, the signals of H-6 and H-6′ are unique to glucose. The α-D anomers of this group have small anomeric coupling constants, whereas the β-D anomers have very large couplings (trans-diaxial).

2. The Lyxose Family

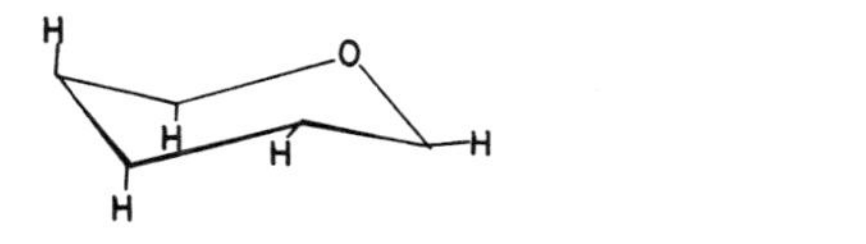

The lyxose family comprises mannose and rhamnose. It is characterized by a small anomeric coupling (the equatorial H-2 precludes any trans-diaxial arrangement). The α-D anomers of this family usually show slightly larger coupling than the β-D anomers. The splitting pattern and coupling constants of the protons on C-3 and C-4 resemble those of the corresponding members of the xylose family.

3. The Arabinose Family

The arabinose family comprises galactose, *N*-acetylgalactosamine, fucose, and arabinose. It is characterized by small H-4 and H-5 couplings and large H-2 and H-3 couplings. The coupling of the anomeric protons in this group is quite similar to that of the xylose family (large for the β-D and small for the α-D anomer).

4. *N*-Acetylneuraminic Acid

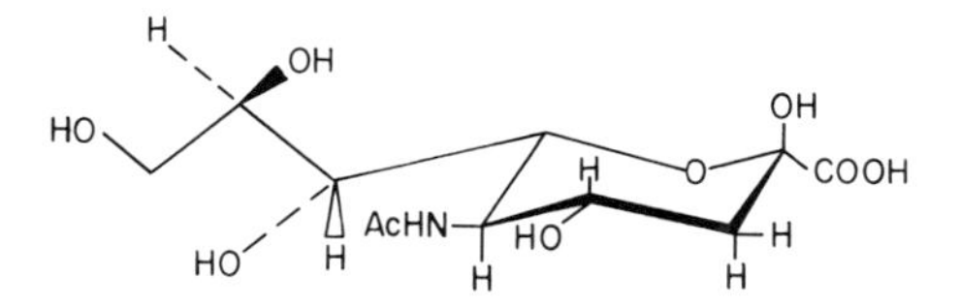

The structure of *N*-acetylneuraminic acid suggests that its ^{1}H-NMR spectrum should reveal, in the high-field region, a singlet due to the methyl protons of the *N*-acetyl group and two multiplets arising from the geminal methylene protons (H-3*a* appears as a triplet at a higher field than the quartet due to H-3*e*). The fact that this compound does not possess any anomeric protons readily distinguishes it from *N*-acetylglucosamine.

Table I summarizes the NMR characteristics of the different monosaccharide components of naturally occurring oligosaccharides, as well as of *N*-acetylneuraminic acid.

B. Examples of Structure Elucidation by ^{1}H-NMR Spectroscopy

To illustrate how the spectral characteristics of the oligosaccharide monomers can be used to elucidate the structure of oligosaccharides, consider the three figures presented next. They depict the ^{1}H-NMR spectra of oligosaccharides isolated from glycoproteins after hydrolyzing the protein moieties. This is why asparagine is linked to some of them through

TABLE I

^{1}H-NMR Characteristics of Natural Oligosaccharide Monomers

D-Monomers	Anomeric proton[a]	H-4[a]	High-field signals
Aldohexoses and 6-deoxyaldohexoses			
Glucose (α)	Doublet (s)	Triplet (l)	—
Glucose (β)	Doublet (l)	Triplet (l)	—
Mannose (α)	Doublet (s)	Triplet (l)	—
Mannose (β)	Singlet	Triplet (l)	—
Galactose (α)	Doublet (s)	Multiplet (s)	—
Galactose (β)	Doublet (l)	Multiplet (s)	—
Fucose (α)	Doublet (s)	Multiplet (s)	Doublet
Fucose (β)	Doublet (l)	Multiplet (s)	Doublet
Rhamnose (α)	Doublet (s)	Triplet (l)	Doublet
Rhamnose (β)	Singlet	Triplet (l)	Doublet
Pentoses			
Arabinose (α)	Doublet (s)	Multiplet (s)	—
Arabinose (β)	Doublet (l)	Multiplet (s)	—
Xylose (α)	Doublet (s)	Triplet (l)	—
Xylose (β)	Doublet (l)	Triplet (l)	—
Amino sugars			
Glucosamine (α-*N*-Ac)	Doublet (s)	Triplet (l)	Singlet
Glucosamine (β-*N*-Ac)	Doublet (l)	Triplet (l)	Singlet
Galactosamine (α-*N*-Ac)	Doublet (s)	Multiplet (s)	Singlet
Galactosamine (β-*N*-Ac)	Doublet (l)	Multiplet (s)	Singlet
N-Acetylneuraminic acid	—	Multiplet	Singlet + two multiplets

[a] l, Large coupling; s, small coupling.

the terminal (amide) nitrogen of this amino acid. As mentioned earlier, glycoprotein oligosaccharides are composed of *N*-acetylneuraminic acid and four monosaccharides (*N*-acetylglucosamine, mannose, galactose, and fucose), which we shall attempt to identify from the spectra.

1. Figure 6 shows the high-resolution ^{1}H-NMR spectra (360 MHz) of *N*-acetyl-β-D-glucosaminyl-(1→N)-L-asparagine (top) and α-D-fucopyranosyl-(1→6)-*N*-acetyl-β-D-glucosaminyl-(1→N)-L-asparagine (bottom). It may be seen that all of the characteristics of the first spectrum (methyl

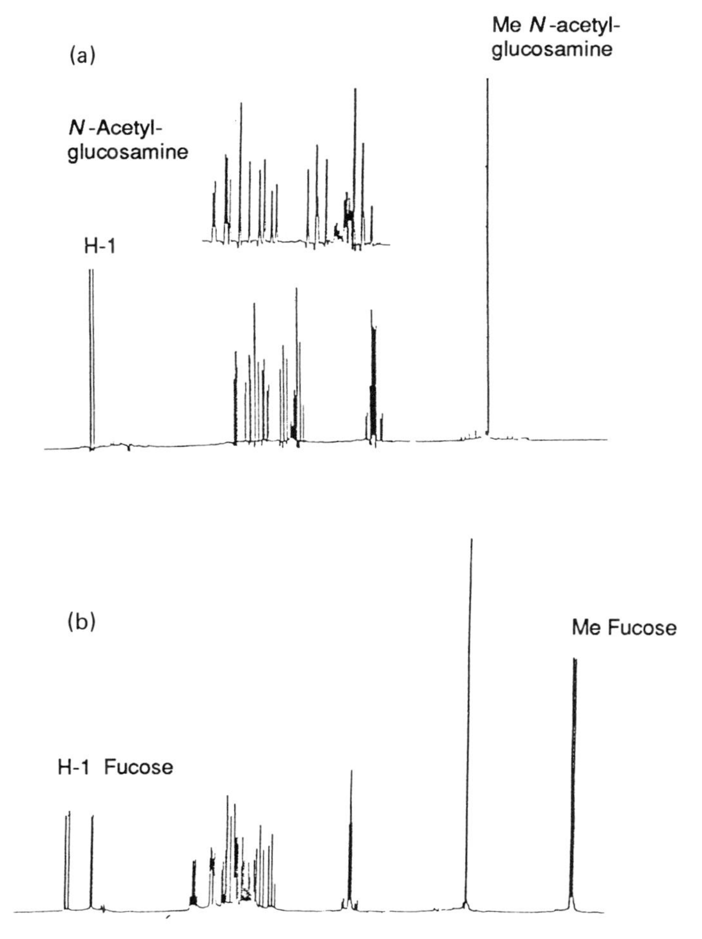

Fig. 6. NMR spectra of *N*-acetyl-β-D-glucosaminyl-asparagine (a) and α-D-fucopyranosyl-(1→6)-*N*-acetylglucosaminyl-asparagine (b).

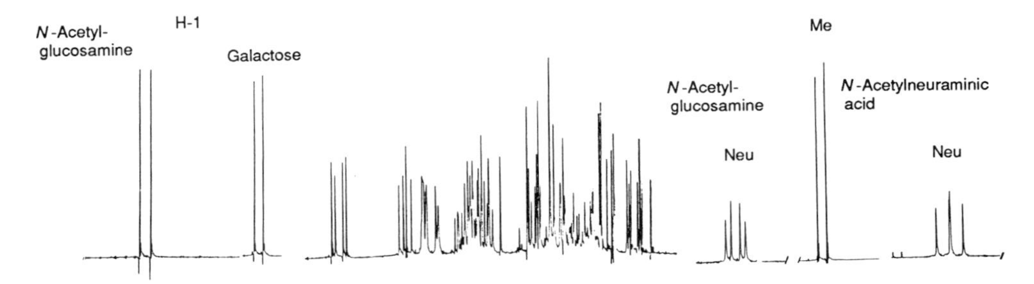

Fig. 7. NMR spectrum of α-*N*-acetylneuraminyl-(2→6)-β-D-galactopyranosyl-(1→4)-*N*-acetyl-β-D-glucosaminyl-asparagine.

protons, large anomeric coupling, and large triplets, for H-3 and H-4) are present in the second spectrum. The latter shows, in addition, the characteristic signals of the α-D-fucosyl group (the methyl doublet, the small anomeric coupling, and the quartet arising from H-5).

2. The next spectrum (Fig. 7) shows a trisaccharide linked to the amide nitrogen of L-asparagine, α-*N*-acetylneuraminyl-(2→6)-β-D-galactopyranosyl-(1→4)-*N*-acetyl-β-D-glucopyranosyl-(1→N)-L-asparagine. The high-field region of the spectrum shows the two *N*-acetyl protons (from *N*-acetylglucosamine and *N*-acetylneuraminic acid), as well as H-3 and H-3′ of *N*-acetylneuraminic acid. The low-field region shows two anomeric protons having a large coupling (β-D configuration), due to H-1 of *N*-acetylglucosamine (at low field) and H-1 of galactose.

3. The last spectrum (Fig. 8) is that of a free trisaccharide, α-D-manno-

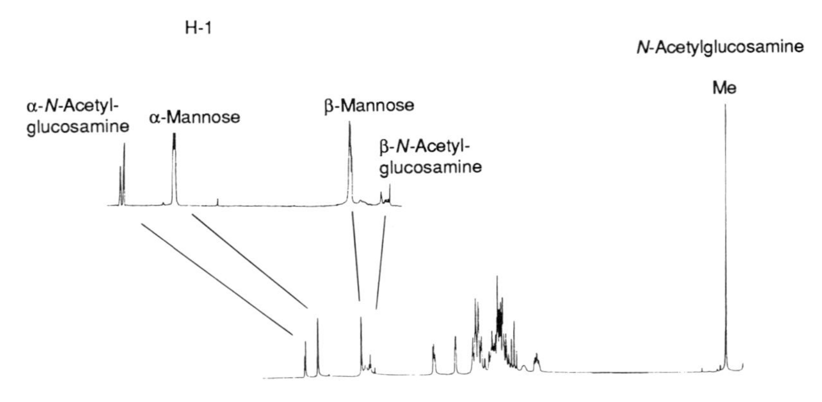

Fig. 8. NMR spectrum of α-D-mannopyranosyl-(1→3)-β-D-mannopyranosyl-(1→4)-*N*-acetylglucosamine. Glucosamine shows signals for both anomers.

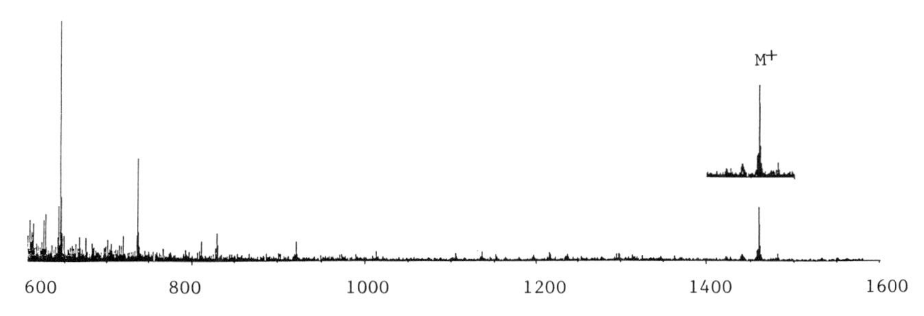

Fig. 9. FAB mass spectrum of 6,6‴-di-*O*-α-D-glucopyranosylcyclomaltoheptaose (M^+1,458).

pyranosyl-(1→3)-β-D-mannopyranosyl-(1→4)-*N*-acetylglucosamine. Four sets of signals are seen, two due to the α and β anomers of the terminal *N*-acetylglucosamine and two due to α-D- and β-D-mannose. It may be seen that the anomeric coupling of the β-D-mannose is smaller than that of the α-D anomer.

C. Mass Spectrometry

Peracetylation or permethylation of oligosaccharides before and after oxidation or reduction gives products that are particularly useful in structure elucidation. The acyclic portion is degraded stepwise during mass spectrometry, giving an insight into the position of branching.

FAB mass spectrometry has found widespread applications in the structure elucidation of oligosaccharides and aminocyclitol antibiotics. This is exemplified by the eludication of the structure of a 6,6‴-di-*O*-α-D-glucopyranosylcyclomaltoheptaose (MW 1,458), whose FAB mass spectrum is depicted in Fig. 9. [For more details see K. Koizumi, T. Utamura, M. Sato, and Y. Yagi, *Carbohydr. Res.* **153,** 55 (1986)].

PROBLEMS

1. Answer the following questions relating to the disaccharide turanose:

(a) What is the reducing power of the disaccharide?
(b) How many moles of periodic acid will its methyl glycoside consume?

(c) What sugar(s) will it yield on oxidation and hydrolysis?
(d) Which methylated sugars will it give if treated with methyl iodide and then hydrolyzed?

2. The clinically used anticoagulant heparin is a polysaccharide made up of 20 tetrasaccharide repeating units $(A\text{–}B\text{–}A\text{–}C)_{20}$. When subjected to partial hydrolysis, it afforded a number of oligosaccharides.
   (a) Give the structures (in letters) of the possible pentasaccharides.
   (b) The structures of the different tetrasaccharides.
   (c) The structures of the different trisaccharides.
   (d) The structures of the possible disaccharides.

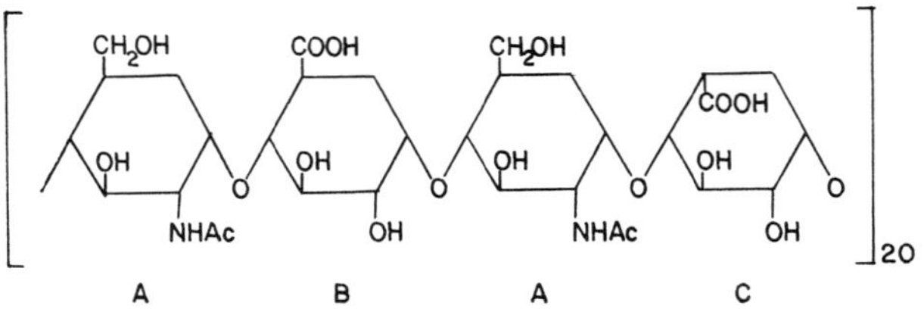

CH_2OH COOH CH_2OH COOH
OH O OH O OH O OH O
NHAc OH NHAc OH
20
A B A C

3. Glycoproteins afford on hydrolysis oligosaccharides composed of five pyranoses: (a) *N*-acetylglucosamine, (b) mannose, (c) galactose, (d) fucose (6-deoxygalactose), and (e) *N*-acetylneuraminic acid (sialic acid). The following NMR spectra depict four oligosaccharides. Assign the proper monomers to the numbered signals (use letters to designate the appropriate monosaccharide).

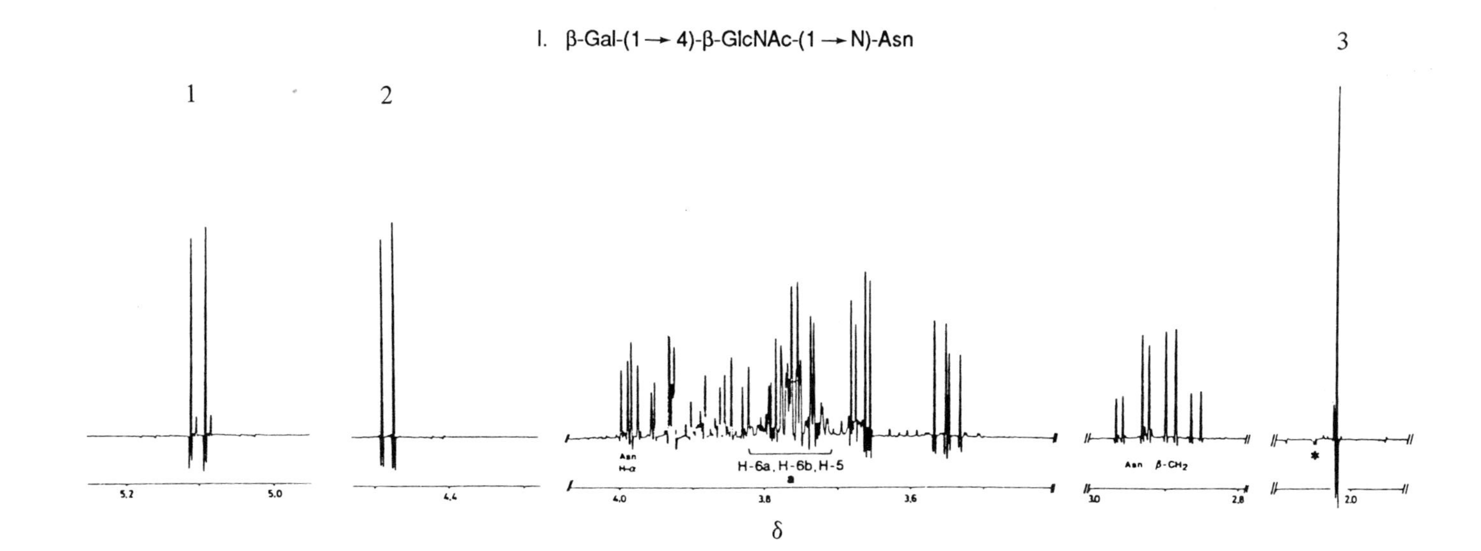
I. β-Gal-(1 → 4)-β-GlcNAc-(1 → N)-Asn
1
2
3
Asn
H-α
H-6a, H-6b, H-5
a
Asn β-CH2
*
5.2
5.0
4.4
4.0
3.8
3.6
3.0
2.8
2.0
δ

II. α-Man-(1 → 2)-α-Man-(1 → 3)-β-Man-(1 → 4)-GlcNAc

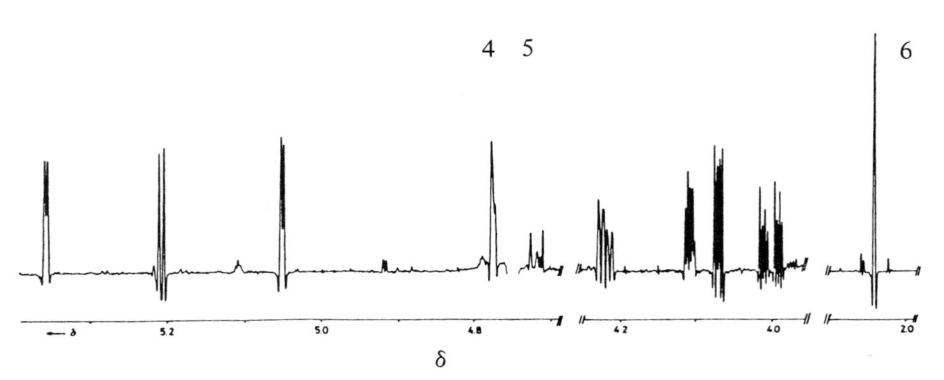

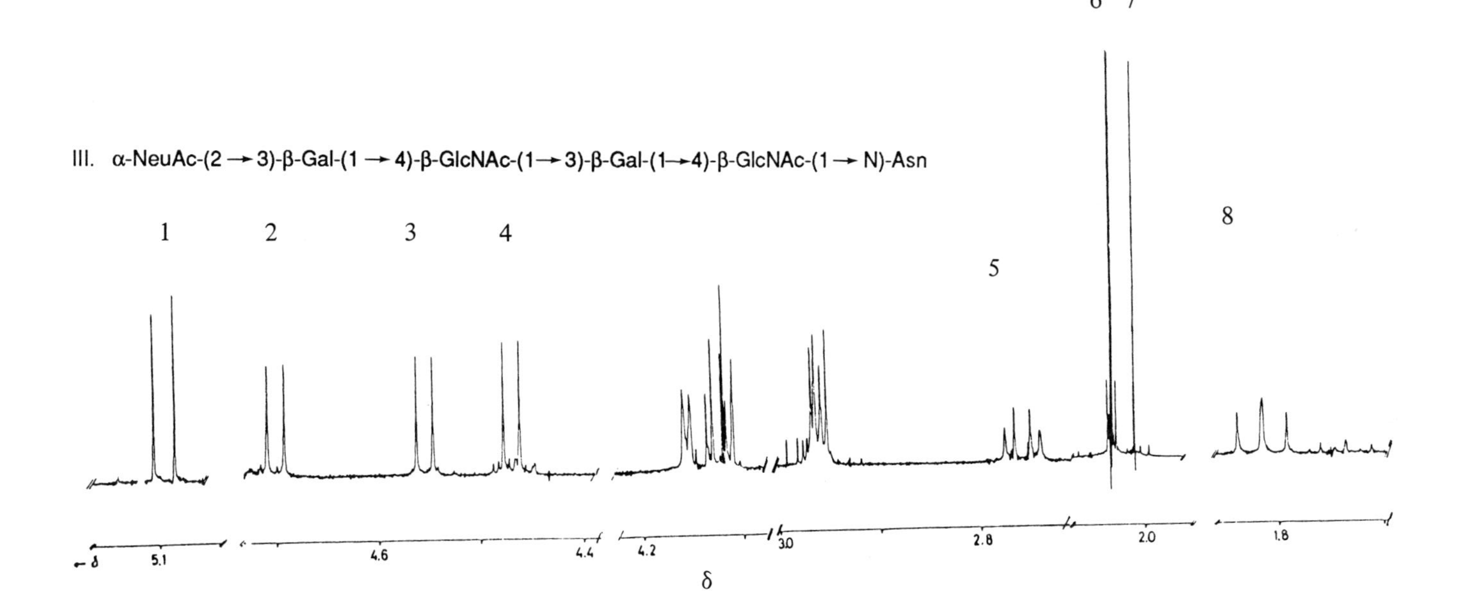
III. α-NeuAc-(2 → 3)-β-Gal-(1 → 4)-β-GlcNAc-(1 → 3)-β-Gal-(1 → 4)-β-GlcNAc-(1 → N)-Asn
1
2
3
4
5
6
7
8
δ
5.1
4.6
4.4
4.2
3.0
2.8
2.0
1.8
δ

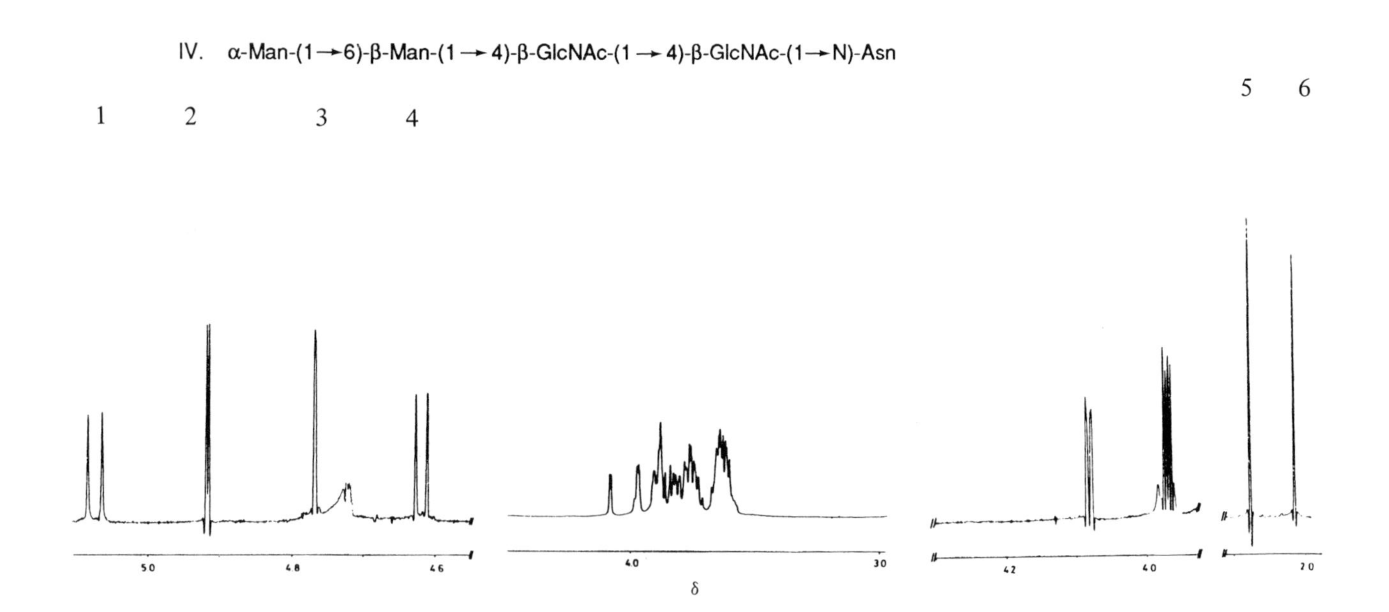
IV. α-Man-(1→6)-β-Man-(1→4)-β-GlcNAc-(1→4)-β-GlcNAc-(1→N)-Asn
1
2
3
4
5
6
5.0
4.8
4.6
4.0
3.0
δ
4.2
4.0
2.0

4. Assign the NMR peaks in the low- and high-field regions of the spectra to the monosaccharide components of the following blood group oligosaccharides.

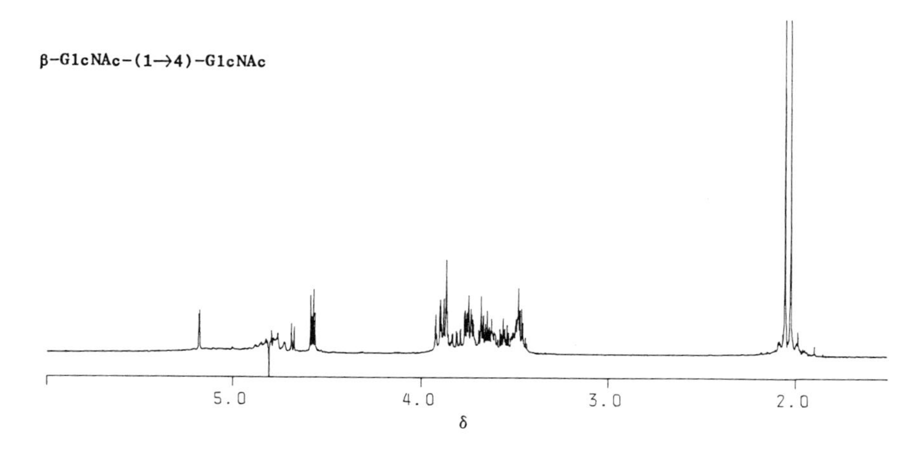

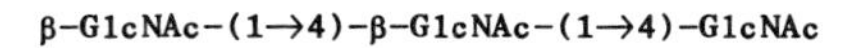

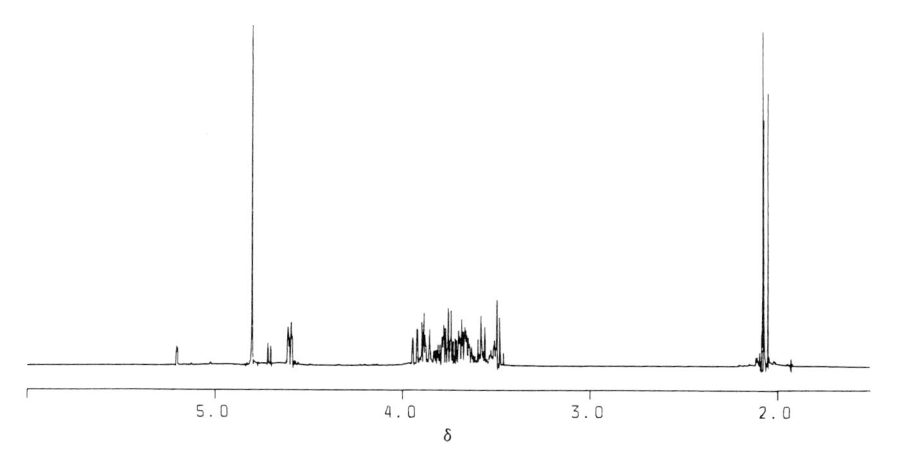

α-Neu-5-Ac-(2→3)-β-Gal-(1→4)-Glc

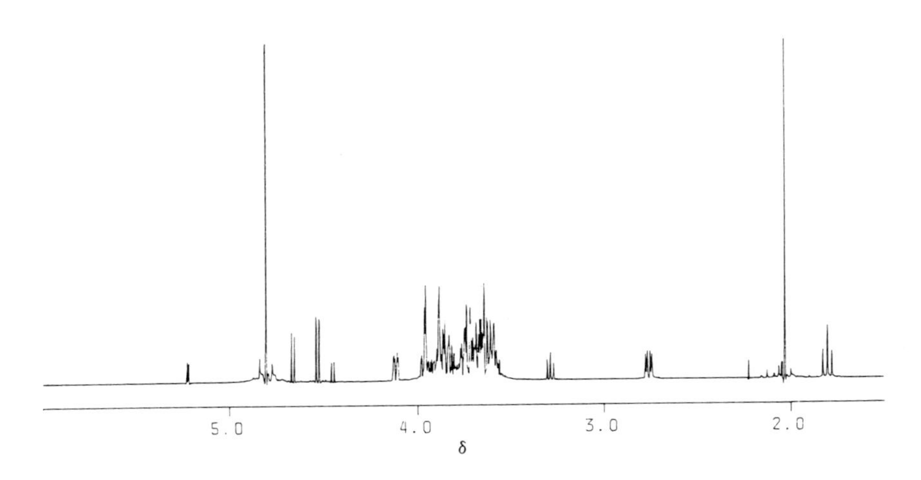

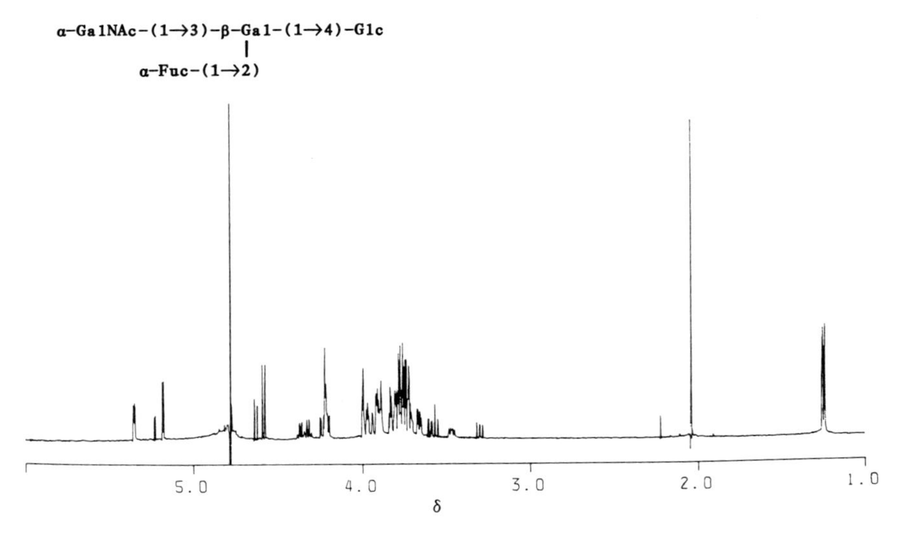

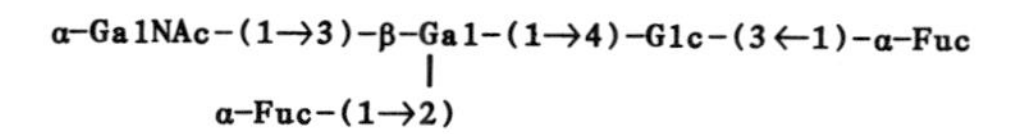

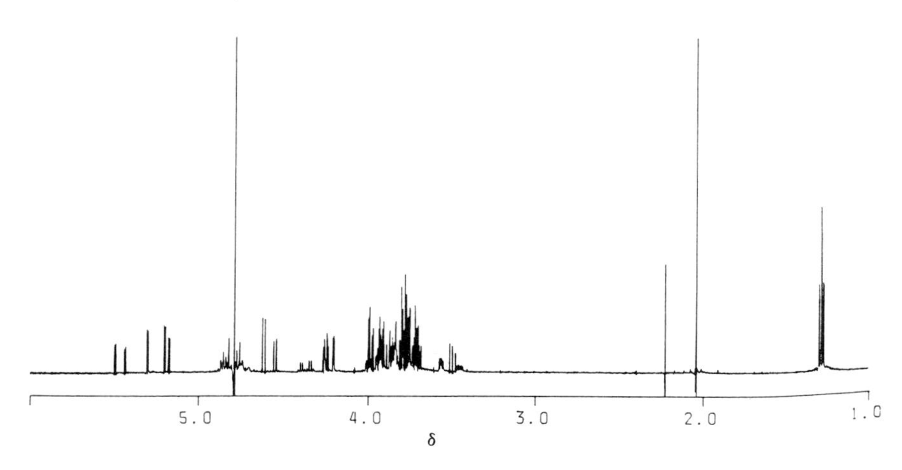

β-GalNAc-(1→3)-α-Gal-(1→4)-β-Gal-(1→4)-Glc

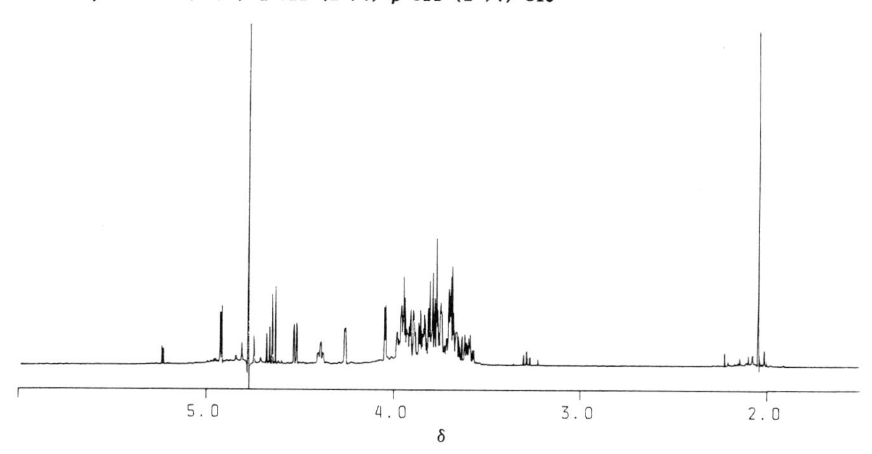

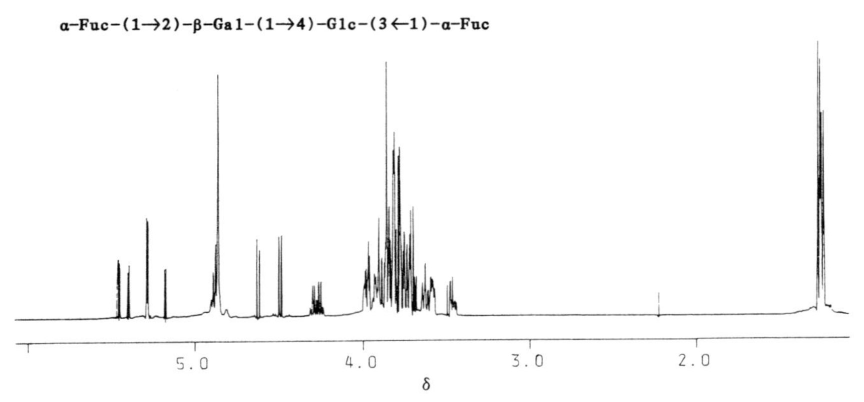

5. In ^{13}C-NMR spectra, substituent effects cause the absorption of the carbon atoms involved in glycosidic linkages or in ester formation to shift to a higher field when the substituents are removed. Show how this rule can be applied to determine the position of a naturally occurring ester group from the ^{13}C-NMR spectra. The upper spectrum represents the natural product before, and the lower after, deacetylation with alkali.

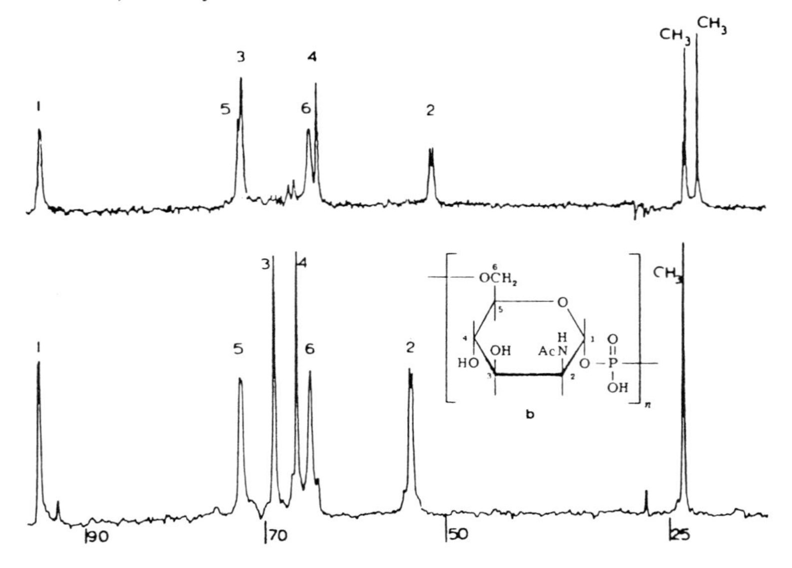

7

Chemical Synthesis and Modifications of Oligosaccharides, Nucleotides, and Aminoglycoside Antibiotics

I. SYNTHESIS OF OLIGOSACCHARIDES

The chemical synthesis of oligosaccharides usually involves reactions between saccharide derivatives that have a good leaving group at the anomeric position, such as a halogen atom or an ester group, and a monosaccharide or an oligosaccharide. In the first case the glycosyl halide is subjected to a Koenigs–Knorr type of reaction, using a halogen acceptor (for example, Ag_2CO_3) as the catalyst, whereas when an ester is used a Helferich type of reaction is carried out with a Lewis acid catalyst.

Whichever glycosidation method is used in oligosaccharide synthesis, two questions must be carefully addressed:

(a) How to ensure that only the oxygen atom in the desired position forms the glycoside bond? This is achieved by ensuring that the desired hydroxyl group is either the most reactive hydroxyl group in the molecule or, better, the only one available for reaction. Because hemiacetal hydroxyl groups are the most reactive hydroxyl groups in cyclic saccharides, it is possible to prepare nonreducing disaccharides by treating glycosyl halides with unprotected monosaccharides. On the other hand, to form reducing disaccharides, it is necessary to protect some or all of the nonreacting hydroxyl groups. Thus, if the desired disaccharide is linked through a primary hydroxyl group [for example, in the case of a (1→6)-linked disaccharide], it is necessary to block (by glycosidation) the anomeric position (because the hemiacetal hydroxyl is more reactive than the primary hydroxyl group). Finally, if more than one primary hydroxyl group is present (as in the case of a ketose), or if one of the secondary hydroxyl groups is to form a glycosidic bond, it is advisable to block all but this particular hydroxyl group to ensure that only the desired glycosidic linkage is formed.

(b) How to ensure that the glycosidic linkage formed is of the desired anomeric configuration? The α or β configuration of a newly formed glycosidic bond is determined, to a large extent, by the blocking groups present in the sugar moiety undergoing nucleophilic attack. Thus, if a participating group (for example, an acetyl or a benzoyl group) is attached to O-2 of the molecule undergoing nucleophilic attack, a Koenigs–Knorr or Helferich type of reaction will yield a glycoside having the trans-1,2 configuration. This isomer is favored because the intermediate cyclic carbonium ion is attacked from the side opposite the ring. This is why β anomers are obtained from D-glucopyranosyl halides (and D-galactopyranosyl halides) and α anomers from D-mannopyranosyl halides.

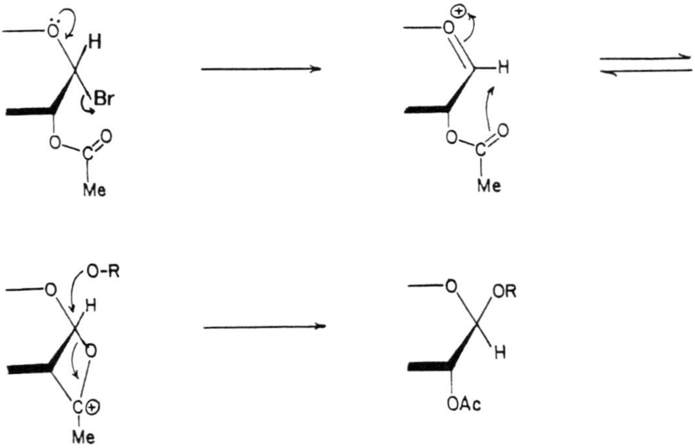

To obtain a cis-1,2 configuration (for example, a glycoside having an α-D-glucopyranosyl, an α-D-galactopyranosyl, or a β-D-mannopyranosyl configuration), OH-2 of the sugar moiety undergoing nucleophilic attack must be protected with a nonparticipating group such as a benzyl group.

In this case, a mixture of α and β anomers is obtained and can be separated by chromatography. The composition of this mixture is influenced by anomeric effects, which favor α-D anomers, and by temperature and duration of reaction, which favor the thermodynamically more stable product, i.e., the one having equatorial substituents [see J.-R. Pougny, J.-C. Jacquinet, M. Nasser, M.-L. Milat, and P. Sinai, *J. Am. Chem. Soc.* **99,** 6762 (1977)].

O H Br OBn → $O^{\oplus}$ $\bar{O}$—R H $\bar{O}$—R OBn → O OR H OBn ; O H OR OBn

The following examples of oligosaccharide syntheses illustrate the ideas just discussed.

Lactose possesses, as its systematic name (4-*O*-β-D-galactopyranosyl-D-glucose) denotes, a β-D glycosidic bond linking C-1 of galactose to O-4

CH_2OAc AcO OAc O Br OAc + EtO H OEt C H—C—O Me_2C O—C—H H—C—OH H—C—O CMe_2 H_2C—O → CH_2OAc AcO OAc O O OAc EtO H OEt C H—C—O Me_2C O—C—H H—C H—C—O CMe_2 H_2C—O

$\xrightarrow[OH^-]{H^+}$ CH_2OH HO OH O O OAc CH_2OH OH O ~OH OH

Lactose

of glucose. It could therefore be synthesized by reaction of a galactopyranosyl halide having participating protecting groups on O-2, with a glucose derivative having all of the hydroxyl groups blocked except for OH-4. Actually, this synthesis was performed by treating 2,3,4,6-tetra-*O*-acetyl-α-D-galactopyranosyl bromide with 2,3:5,6-di-*O*-isopropylidene-D-glucose diethyl acetal, under Koenigs–Knorr conditions, and then hydrolyzing the isopropylidene groups with acid and the acetyl groups with base.

Now consider the synthesis of raffinose, which is *O*-α-D-galactopyranosyl-(1→6)-α-D-glucopyranosyl β-D-fructofuranoside. It is evident that the galactopyranosyl halide needed for this synthesis must be protected by a nonparticipating group in order to obtain the desired α-D-galactopyranosyl linkage (cis-1,2 configuration). Furthermore, because the adduct (sucrose) is a nonreducing disaccharide that possesses three primary hydroxyl groups (all of which are available for attack on the carbonium ion),

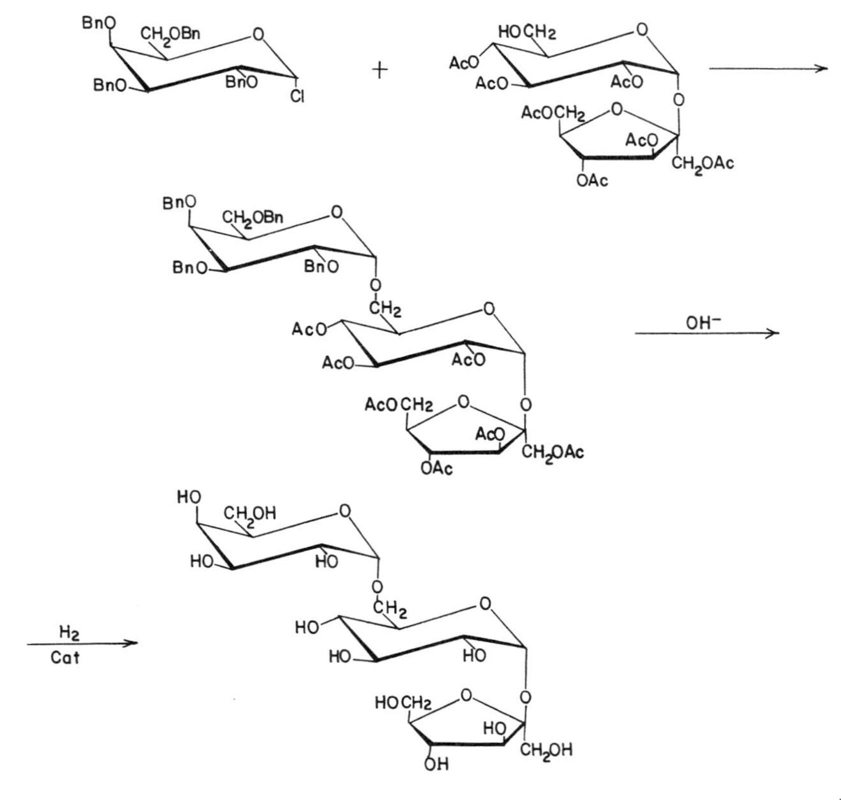

[T. Suami, T. Otake, T. Nishimura, and T. Ikeda, *Carbohydr. Res.* **26,** 234 (1973).]

it is necessary to block at least two of them to ensure that the desired primary hydroxyl group, namely the one attached to C-6 of the glucopyranose moiety, is the one that reacts with the galactopyranosyl halide. Experimentally, raffinose was synthesized by treating, under Koenigs–Knorr conditions, a benzyl-protected galactopyranosyl halide (tetra-*O*-benzyl-α-D-galactopyranosyl chloride) with a sucrose derivative (2,3,4,1′,3′,4′,6′-hepta-*O*-acetylsucrose) having ester groups replacing all of the hydroxyl groups, except OH-6 of the glucopyranosyl moiety. This sequence of reactions constitutes a total synthesis, because sucrose had been synthesized previously. (Several syntheses of this sugar have been reported, first by Lemieux and Huber and then by Tsuchida, Fletcher, Fraser-Reid, and Ogawa and their co-workers.)

A. Block Synthesis

Although it is possible to achieve the synthesis of higher oligosaccharides by the addition of one monosaccharide at a time, it is often advantageous to add the oligosaccharide components of a large oligomer in blocks of two or more monosaccharides (block synthesis). Thus, the

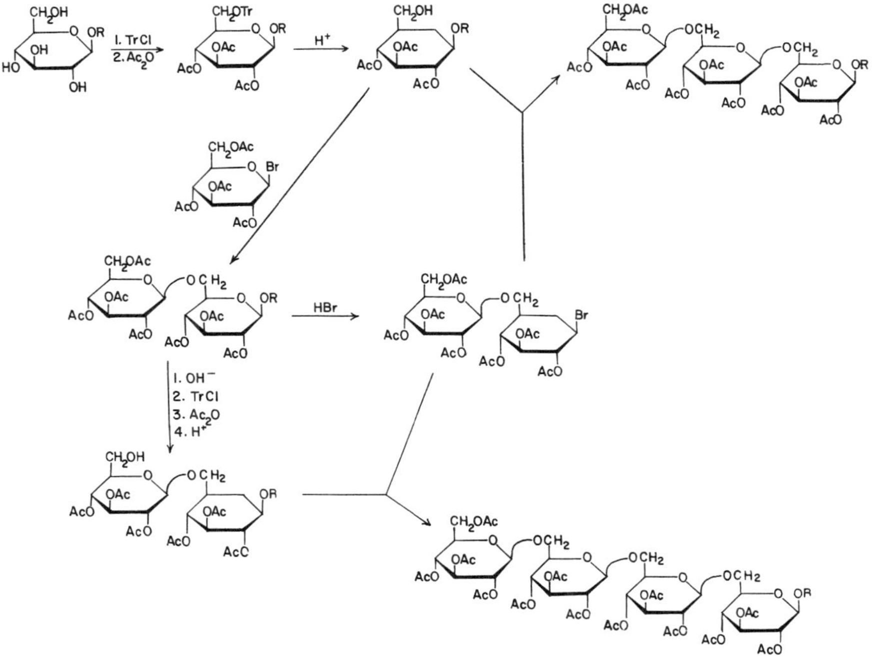

[K. Takiura, M. Yamamoto, Y. Miyaji, H. Takai, and H. Yuki, *Chem. Pharm. Bull.* **22,** 2451 (1974).]

trisaccharide and tetrasaccharide shown below, can be prepared by treating a disaccharide glycosyl halide with a suitably protected mono- or disaccharide. It is evident that a second round of reactions would yield penta- and hexasaccharides.

Temporary protecting groups proved particularly effective in this respect. For example, the allyl glycosides were effectively used to block the anomeric position of the terminal saccharide for ultimate removal with base when the synthesis is complete [see M. A. Nashed and L. Anderson, *Carbohydr. Res.* **56,** 419 (1977)].

B. Synthesis on Polymer Support

Another interesting approach to the chemical and enzymatic synthesis of oligosaccharides is the use of a stationary polymer support to retain substrates (or enzymes) inside a reactor (usually a column), while successive reagents are passed through the polymer and are then washed out of the reactor. The advantage of this method is that it provides an efficient

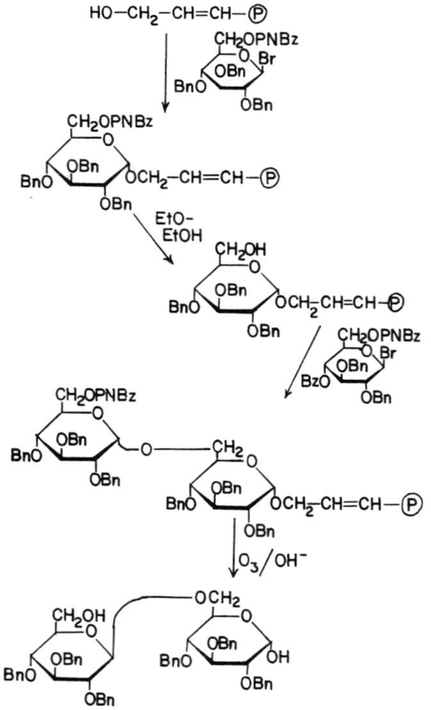

[J. M. Frechet and C. Schuerch, *J. Am. Chem. Soc.* **93,** 492 (1971).]

way to carry out successive reactions without loss of the material attached to the column. For industrial enzymatic reactions involving costly enzymes, retaining the enzymes inside the reactor would seem quite attractive. Also advantageous is the stepwise chemical synthesis of a homologous oligosaccharide, such as the one just shown. The first monomer is linked to a synthetic polymer (usually a substituted polystyrene), and successive monomer units are attached to it by Koenigs–Knorr reactions. When the last monomer has been attached, the oligomer formed is removed under mild conditions that will not cause depolymerization.

An active area of research involves the linkage of oligosaccharide 1-phosphates to naturally occurring alcohols of oligomeric alkenes, such as dolichol, to yield lipids and lipid intermediates [see C. D. Warren, M.-L. M. Lat, C. Augé, and R. W. Jeanloz, *Carbohydr. Res.* **126,** 61 (1984)].

II. MODIFICATION OF OLIGOSACCHARIDES

Chemical modifications of oligosaccharides usually involve the reactive primary hydroxyl group. As the degree of polymerization of an oligosaccharide increases, directing a substituent in a particular moiety becomes more difficult. In a disaccharide, it is possible to direct a substituent toward the terminal nonreducing saccharide moiety by blocking the primary hydroxyl group of the reducing moiety with a 1,6-anhydro ring. The latter is introduced by treating the disaccharide glycoside (usually a phenyl glycoside) with a base. Two examples of selective reactions involving the terminal ring of a disaccharide are depicted.

(a) Introduction of a carboxylic group by oxidation of the primary hydroxyl group of benzyl β-cellobioside with oxygen in the presence of palladium on charcoal occurs preferentially at the terminal ring. However, greater selectivity may be achieved if the 1,6-anhydride is formed first.

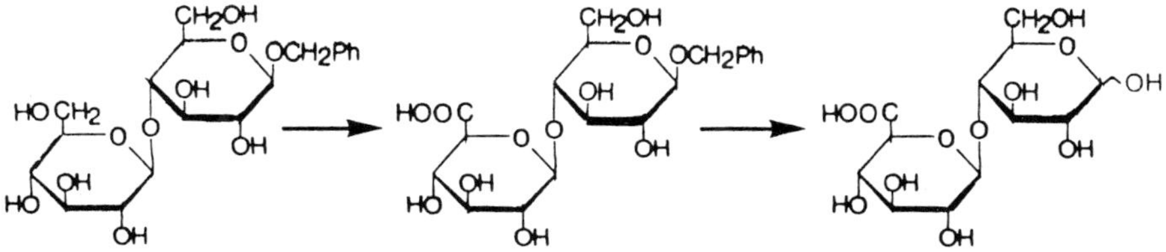

HOCH2 ... CH2OH ... OCH2Ph → HOOC ... CH2OH ... OCH2Ph → HOOC ... CH2OH ... OH

[G. Jayne and W. Demming, *Chem. Ber.* **93,** 365 (1960).]

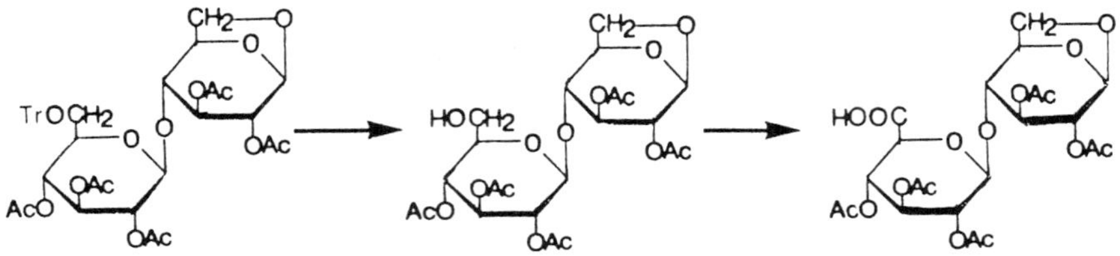

[B. Lindberg and L. Shellby, *Acta Chem. Scand.* **14,** 1051 (1960).]

(b) Introduction of an azido group onto the C-6′ of lactose can also be achieved by tosylating the 1,6-anhydride of cellobiose, tritylating this, and fully acetylating the product. When the trityl group is removed by acid, migration of the acetyl group attached to O-4 occurs, yielding a

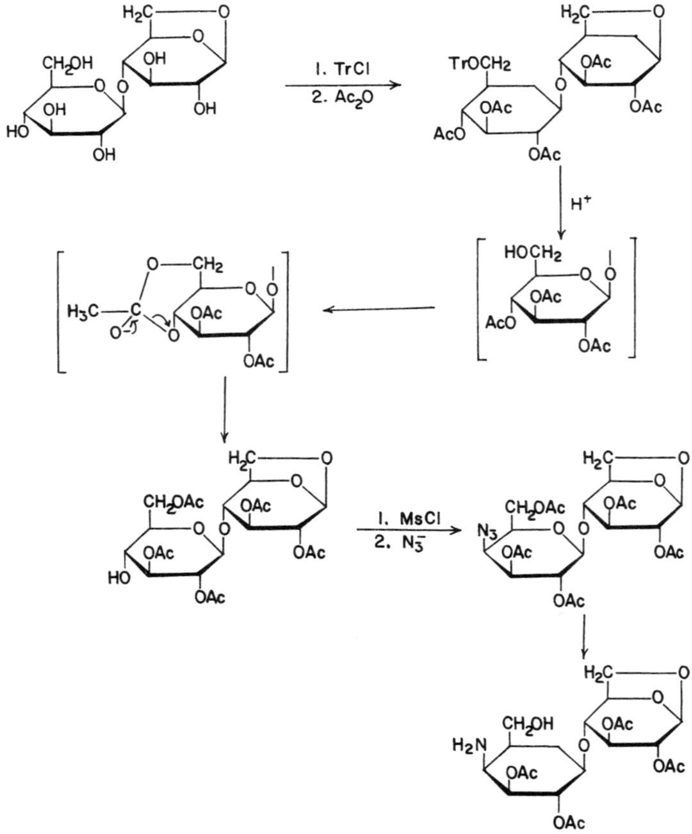

[S. Tejima, Y. Okamori, and M. Haga, *Chem. Pharm. Bull.* **21,** 2538 (1973).]

6-*O*-acetyl-4-hydroxy derivative. Mesylating at O-4 and displacing the leaving group with azide gives the 4-azido derivative of lactose. The azido group can then be reduced to an amino group.

A. Oligosaccharide Surfactants

Of considerable importance as surfactants are the fatty acid esters of disaccharides. An example of these is sucrose mono-oleate, which has a hydrophilic–lipophilic balance (HLB) of 15, ideal for its use as a mild shampoo. Such esters are obtained by treating the disaccharide in *N-N*-dimethylformamide with the methyl ester of the fatty acid.

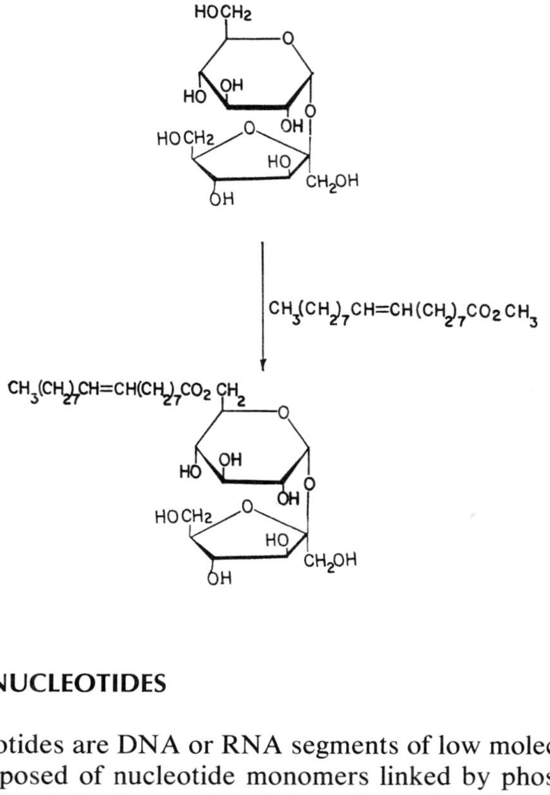

$$CH_3(CH_2)_7CH{=}CH(CH_2)_7CO_2CH_3$$

III. OLIGONUCLEOTIDES

Oligonucleotides are DNA or RNA segments of low molecular weight; they are composed of nucleotide monomers linked by phosphoric ester bridges spanning C-5′ of one unit and C-3′ of the other. Synthetic oligonucleotides have been widely used in the study of DNA, in protein biosynthesis, and in induced mutagenesis. They were instrumental in deciphering the genetic code, and they have since been used to introduce mismatches in specific sites of DNA strands to induce specific changes in enzyme proteins. This has enabled biochemists to replace specific amino acids by others and study the effect on the catalytic properties of enzymes. In this chapter, two methods of synthesis of oligonucleotides will be discussed: the block synthesis approach, and stepwise synthesis on a polymer support. The latter technique has been developed to a high degree of reliability, and programmable DNA synthesizers (analogous to the automatic protein synthesizers) are commercially available.

A. Synthesis of Oligonucleotides

Both oligodeoxyribonucleotides (DNA-type oligomers) and oligoribonucleotides (RNA-type oligomers) have been synthesized, but the first have attracted most of the attention of chemists because of their biological importance and their relative ease of preparation. These characteristics become evident when the structures of the two types of nucleotide monomers are examined. Thus, the DNA monomers have only one free hydroxyl group (on C-3′), which, when esterified by the phosphate group of another monomer, will form a phosphate bridge in the desired position (linking C-3′ of one nucleotide to C-5′ of the other), whereas ribonucleotides have two free hydroxyl groups (one on C-2′ and one on C-3′), which necessitates that the 2′-OH group be protected to prevent the formation of the undesired 2′ → 5′ phosphate bridge. In this chapter, only the synthesis of oligodeoxyribonucleotides will be discussed. [For a review on the synthesis of nucleosides see K. A. Watanabe, D. H. Hollenberg, and J. J. Fox, *J. Carbohydr., Nucleosides, Nucleotides* **1,** 1 (1974).]

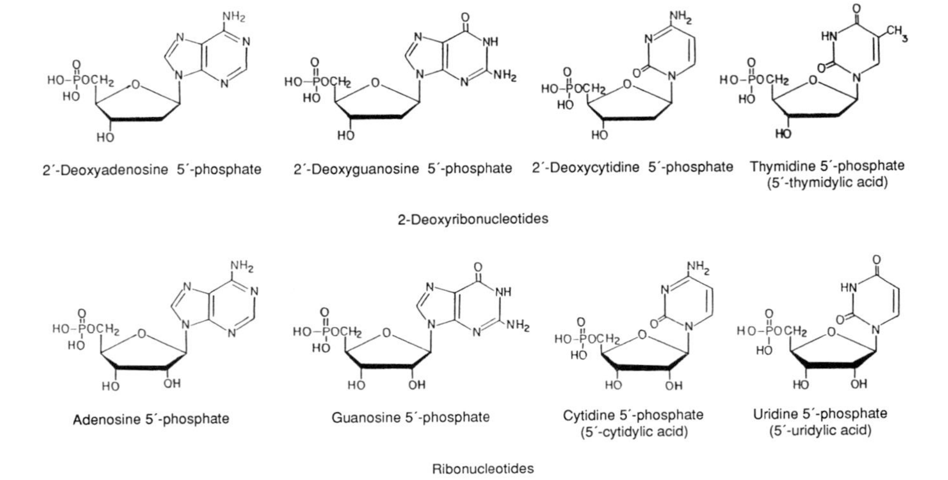

B. Block Synthesis of Oligonucleotides

An oligonucleotide chain may be terminally phosphated (esterified) on C-5′ or C-3′. Oligomers of the first type are prepared from natural nucleotides by treating an acetylated nucleotide (one having a phosphate group on C-5′ and an acetyl group on O-3′) with a nucleotide or an oligonucleotide having a blocked phosphate group on C-5′ and a free hydroxyl group in position 3′. The second type, those terminally phosphated at C-3′, are usually prepared from nucleosides phosphated at C-3′ (which are not available commercially). The monomers are acetylated on O-5′ and treated with other monomers having free hydroxyl groups on C-5′ and a blocked phosphate group on C-3′ to obtain the desired oligonucleotide.

In both cases, it is necessary to block the phosphate groups to prevent the formation of anhydrides (pyrophosphates). This is done by treating the nucleotide with 3-hydroxypropanonitrile to obtain an ester removable with alkali (β elimination). Other phosphate-blocking groups include $PhCH_2—O—(CH_2)_2—NH_2$, which forms an amide hydrolyzable by acids.

The coupling of the two nucleotide moieties involves formation of a phosphoric diester from a monoester (or a triester from a diester). The reaction is catalyzed by such condensing agents as dicyclohexylcarbodiimide (DCC), a reagent extensively used in peptide synthesis, or arylsulfonyl chlorides (for example, 2,4,6-trimethylphenyl- or 2,4,6-triisopropylphenylsulfonyl chloride).

All nucleotides possessing primary amino groups (three of the four bases present in DNA and RNA nucleotides), namely adenylic, guanylic, and cytidylic acid, must have their amino groups protected to prevent the formation of amide ester bridges (instead of diester phosphate bridges). The primary amino group of adenylic acid is usually protected by a benzoyl (or *p*-methoxybenzoyl) group. This is achieved by treating the nucleotide with an excess of benzoyl chloride and removing the undesired benzoyl groups with alkali. Guanylic acid is usually protected by reaction with *N,N*-dimethylformamide diethyl acetal to give a readily removable Schiff base. Finally, cytidylic acid is protected by formation of a carbamate.

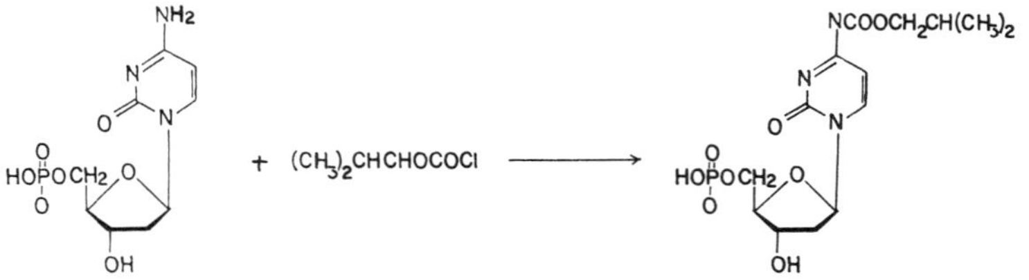

The synthesis of an oligodeoxynucleotide having a terminal phosphate group at C-5′ may be exemplified by the following scheme.

a. Pretreatment of the nucleotide monomers:

1. Nucleotides having primary amino groups must have these groups blocked with the appropriate groups already discussed.

2. The terminal nucleotide must have its phosphate group blocked (for example, by reaction with 3-hydroxypropanonitrile), leaving its 3′-hydroxyl group free.

3. All subsequent nucleotides are acetylated at O-3′, and their 5′-phosphate groups are left free.

b. Coupling reactions:

1. First coupling. The first and second nucleotides are allowed to react in the presence of a condensing agent (DCC). The dimer formed is deacetylated with a base, which causes the removal (by β elimination) of the phosphate protecting group and yields the dimer.

2. Second coupling. The dimer must first react with 3-hydroxypropanonitrile, and the product acetylated. The acetate is then treated with the third nucleotide plus DCC. Finally, the product is treated with alkali to obtain the trimer (a codon).

3. Third coupling. The previous trimer is treated with 3-hydroxypropanonitrile and then with another trimer (preacetylated) in the presence of DCC. Finally, the product is treated with alkali to obtain the hexamer.

4. Fourth coupling. To link two hexamers, the same procedure is repeated. The resulting hexamer is treated with 3-hydroxypropanonitrile, the product coupled with an acetylated hexamer, and the product deblocked with base.

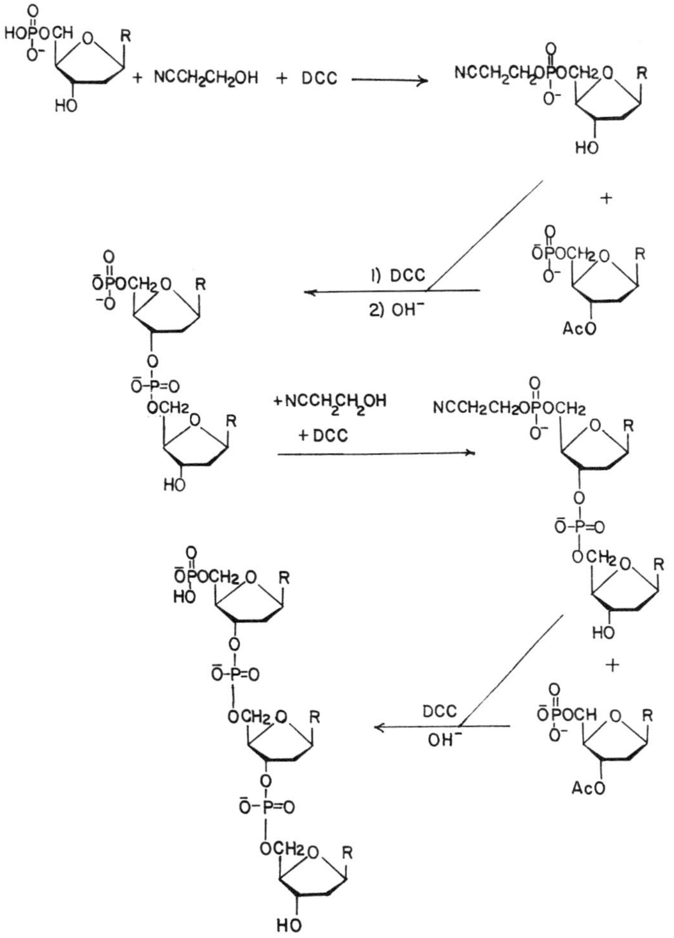

The synthesis of an oligonucleotide having a terminal phosphate on C-3′ may be carried out by coupling two nucleotides having aryl phosphate groups in position 3′. The first nucleotide would have an acid-labile *p*,*p*′-dimethoxytrityl group on position 5′. The second would have a cyanoethyl group, which is base labile, protecting its phosphate group. Deblocking the dimer with base would yield a dimer which can later be coupled with another (a) monomer to obtain a trimer or (b) a dimer having a cyanoethyl group on the phosphate group to obtain a tetramer. Repeating the process produces hexamers and octamers.

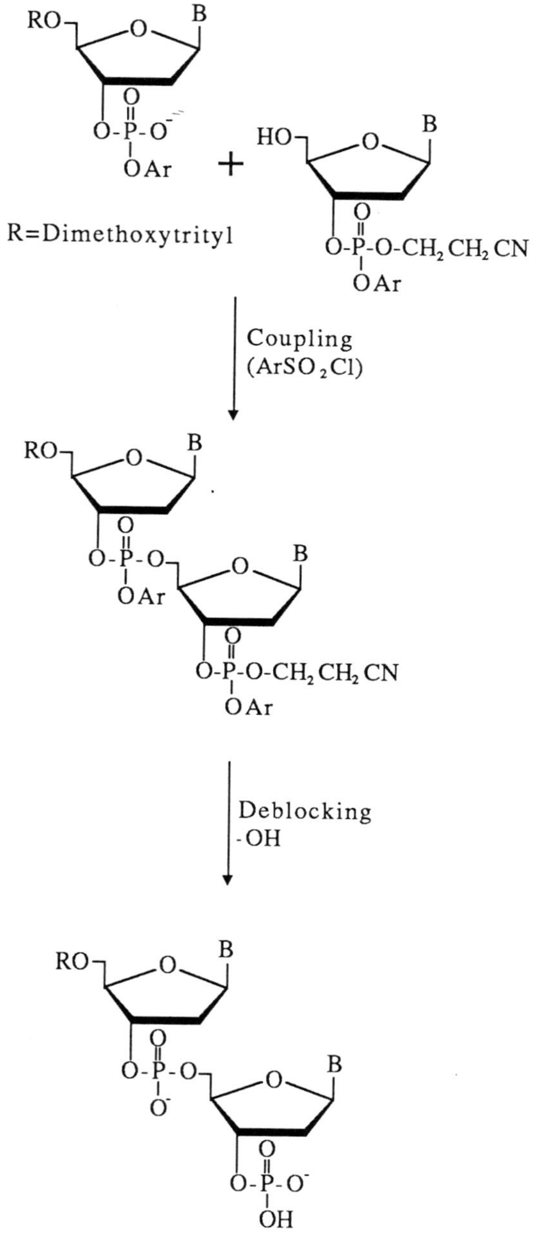

RO B O

O
O-P-O⁻
OAr

\+

HO B O

O
O-P-O-CH_2CH_2CN
OAr

R=Dimethoxytrityl

Coupling
($ArSO_2Cl$)

RO B O

O
O-P-O B O
OAr

O
O-P-O-CH_2CH_2CN
OAr

Deblocking
-OH

RO B O

O
O-P-O B O
O⁻

O
O-P-O⁻
OH

C. Stepwise Synthesis on Polymer Support

1. General Considerations

Syntheses carried out on a polymer support offer many advantages, particularly if the steps are repetitive. This is certainly the case with oligonucleotides, as there are only four monomers, which are coupled in the desired order, using identical procedures. In designing a synthesis that is suitable for automation, it is important to decide which part of the nucleotide molecule will be attached to the polymer. The answer is definitely the phosphate group in the case of terminally 5′-phosphated oligomers and C-5′ of the sugar in the case of terminally 3′-phosphated oligomers. Attachment through the bases is not desirable because it cannot be used for all nucleotides. Some bases lack amino groups that can bind to the resin. Examples of phosphate-linked nucleotides and sugar-linked nucleotides are shown on the next page. At present the "phosphite" method is more commonly used, and is also used in commercial automatic polynucleotide synthesizers.

2. Nucleoside Analogs as Antitumor and Antiviral Agents

Modified nucleosides have been used to offset the rapid replication of the DNA of cancer cells the RNA of viruses. Modified purine and pyrimidine bases and their nucleosides have both been used effectively. Examples of modified purines used as antitumor agents are 6-thiopurine and its D-ribofuranosyl derivative, and an example of an antiviral pyrimidine is 3′-azido-3′-deoxythymidine (AZT), which is used in treating acquired immunodeficiency syndrome (AIDS).

To synthesize nucleoside analogues having the furanosyl ring linked to the nitrogen of the base, the same technique used for the synthesis of

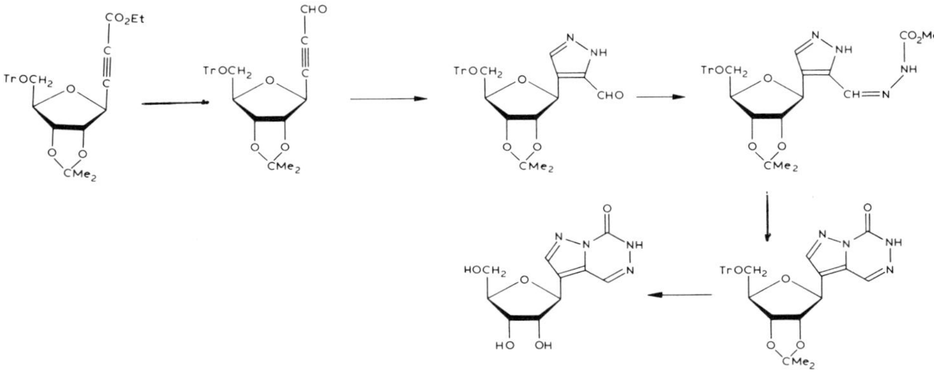

[D. L. Swartz and H. El Khadem, *Carbohydr. Res.* **112**, C1 (1983).]

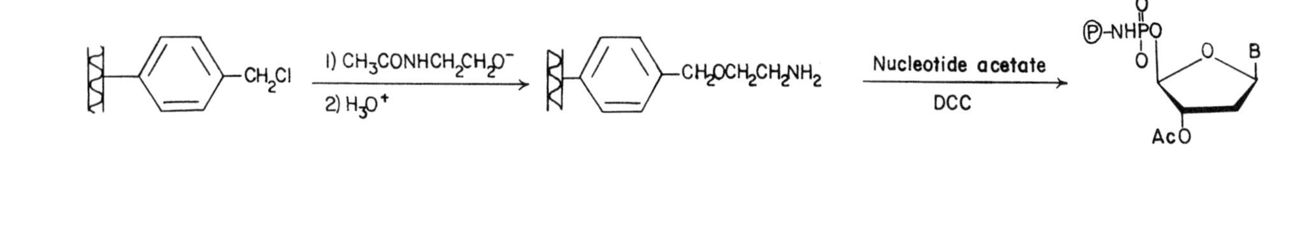
-CH2Cl
1) CH3CONHCH2CH2O⁻
2) H3O+
-CH2OCH2CH2NH2
Nucleotide acetate
DCC
P-NHPO
AcO
B

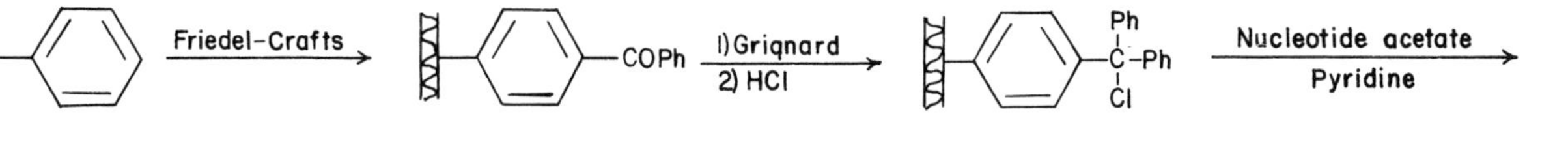
Friedel-Crafts
-COPh
1) Grignard
2) HCl
Ph
C-Ph
Cl
Nucleotide acetate
Pyridine

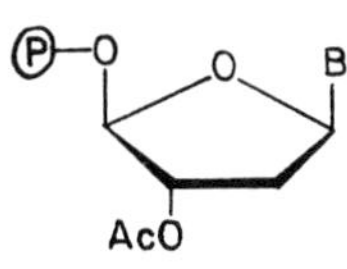
P-O
B
AcO

natural nucleosides may be used. If, on the other hand, the glycosidic linkage involves one of the carbon atoms of the base, then the heterocyclic ring must be closed after attachment to the furanose ring. An example of the synthesis of such a C-nucleoside is depicted on the previous page. The acetylinic group is attached to the ring by a Grignard reaction via a glycosyl halide. Then cycloaddition with diazomethane on the β isomer closes the five-membered ring. Finally, reaction with the appropriate hydrazine closes the six-membered ring.

IV. SYNTHESIS OF AMINOGLYCOSIDE ANTIBIOTICS

Aminoglycosides constitute a group of effective antibiotics composed of aminated oligosaccharides linked to aminocyclitols. The first member of this group of antibiotics, namely streptomycin, was isolated by Waks-

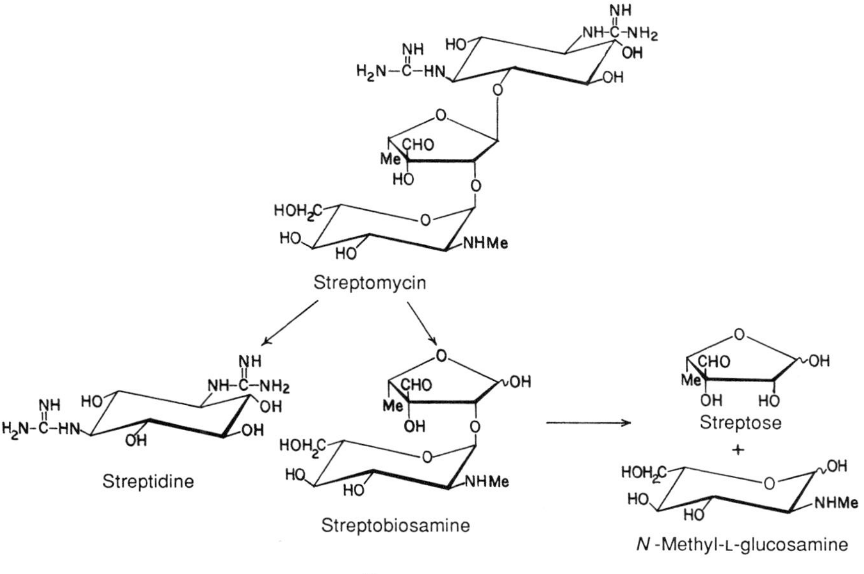

Streptomycin

Streptidine

Streptobiosamine

Streptose

+

N-Methyl-L-glucosamine

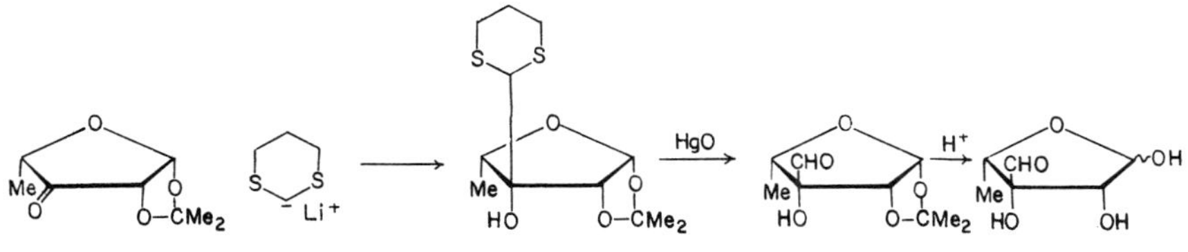

Streptose

[H. Paulsen, V. Sinnwell, and P. Stadtler, *Angew. Chem., Int. Ed. Engl.* **11,** 149 (1972).]

man in 1944 from soil bacteria. Its structure was determined by partial hydrolysis, which yielded an aminocyclitol, streptidine, and a disaccharide, streptobiosamine, composed of α-L-glucosamine and the branched sugar streptose. The latter was synthesized by treating the L-*threo* derivative shown with 2-lithio-1,3-dithiane and hydrolyzing the adduct.

The synthesis of the oligosaccharide moiety of aminoglycosides from their monomers by Koenigs–Knorr or Helferich methods offers some difficulty with anomeric control. The amino sugar monomers are prepared

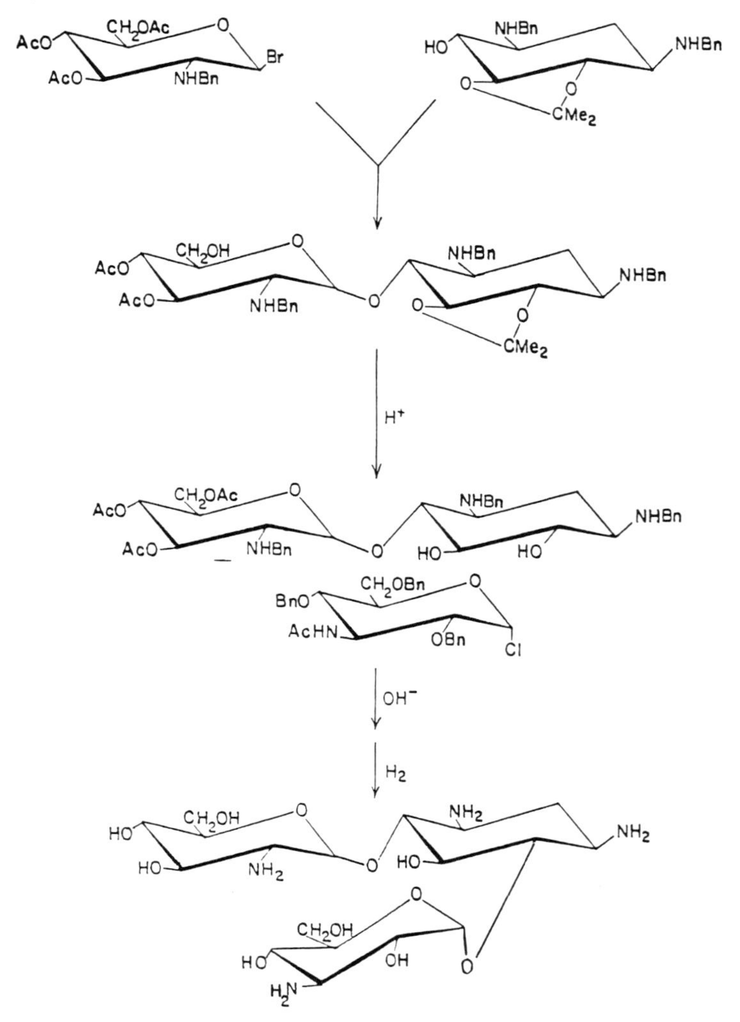

[S. Umezawa, S. Koto, K. Tatsua, and T. Tsumara, *J. Antibiot.* **21,** 162 (1969).]

by introducing the desired amino groups into monosaccharide molecules by standard methods, for example, reduction of oximes or azides. The synthesis of the required branched sugars involves reaction of carbonyl groups with carbanions, in the case of streptose. A potential difficulty in aminoglycoside synthesis may arise during the coupling of the saccharides with the aminocyclitol. If more than one monosaccharide is to be coupled to an aminocyclitol, it is difficult to ensure that all the saccharides are attached to the desired hydroxyl group. But, if only one saccharide is to be attached to a cyclitol having three hydroxyl groups, the problem may be solved by forming a cyclic acetal. Two of the three hydroxyl groups in 2-deoxystreptamine are equivalent, so only one isopropylidene derivative (having the desired position free) can be formed. This is made use of in linking the first saccharide residue of the antibiotic kanamycin C. To introduce the second saccharide residue, the isopropylidene group is removed, exposing two hydroxyl groups, so that reaction with the second halide can afford two isomers. Fortunately, the desired product, kanamycin C, is the major reaction product.

PROBLEMS

1. Starting with the monosaccharides fucose (6-deoxy-D-galactose) and *N*-acetylglucosamine (2-acetamido-2-deoxy-D-glucose), show how you would synthesize the disaccharide β-Fuc(1→6)-β-GlcNAc.
2. Give retrosynthetic schemes showing how the anthracycline antibiotic shown could be prepared from available synthons (the aglycon and commercially available sugars).

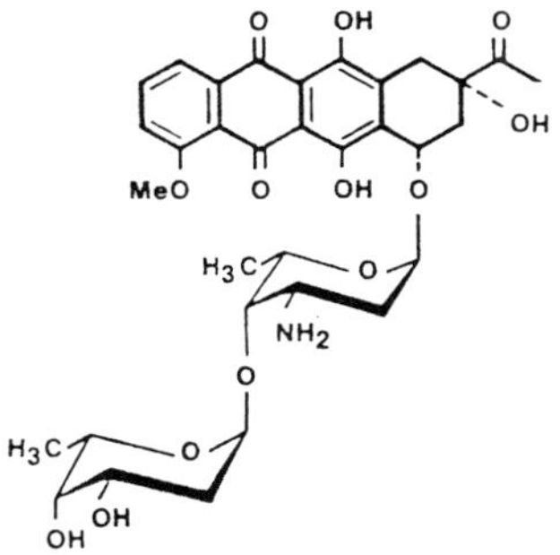

3. Show how you would obtain a 3′ terminally phosphated oligonucleotide, starting with natural (5′-phosphated) nucleotides.

Bibliography

RECOMMENDED TEXTBOOKS

1. "The Carbohydrates," edited by W. Pigman and D. Horton, Vol. IA. Academic Press, New York, 1972.
2. "The Carbohydrates," edited by W. Pigman and D. Horton, Vol. IB. Academic Press, New York, 1980.
3. "The Carbohydrates," edited by W. Pigman and D. Horton, Vols. IIA and IIB. Academic Press, New York. 1970.
4. "Total Synthesis of Natural Products," by S. Hanessian. Pergamon, New York, 1983.
5. *Advances in Carbohydrate Chemistry and Biochemistry,* edited by R. S. Tipson and D. Horton, from Vol. 29 (1974) to Vol. 45 (1987). Academic Press, New York.
6. *Methods in Carbohydrate Chemistry,* edited by R. L. Whistler and J. N. BeMiller, Vols. I, II, and VI. Academic Press, New York.
7. "Stereochemistry of Carbohydrates," by J. F. Stoddart. Wiley (Interscience), New York, 1971.

RECOMMENDED REVIEW ARTICLES

The following review articles are highly recommended for supplemental reading. The references are arranged by chapter and are listed in the same order as the topics discussed.

Chapter 2: Structure, Configuration, and Conformation of Monosaccharides

1. Structure and stereochemistry of the monosaccharides, by W. Pigman and D. Horton, *in* "The Carbohydrates," Vol. IA, p. 1 (1972).

2. Conformations of sugars, by S. J. Angyal, *in* "The Carbohydrates," Vol. IA, p. 195 (1972).
3. The composition of reducing sugars in solution, by S. J. Angyal, *Adv. Carbohydr. Chem. Biochem.* **42,** 15 (1984).

Chapter 3: Nomenclature

1. Instructions to authors, *Carbohydr. Res.* **132,** 185 (1984).

Chapter 4: Physical Properties Used in Structure Elucidation

1. High-resolution nuclear magnetic resonance spectroscopy, by L. D. Hall, *in* "The Carbohydrates," Vol. IB, p. 1299 (1980).
2. Carbon-13 nuclear magnetic resonance spectroscopy of monosaccharides, by K. Bock and C. Pedersen, *Adv. Carbohydr. Chem. Biochem.* **41,** 27 (1983).
3. Infrared spectroscopy, by R. S. Tipson and F. S. Parker, *in* "The Carbohydrates," Vol. IB, p. 1394 (1980).
4. Electronic (ultraviolet) spectroscopy, by H. S. El Khadem and F. S. Parker, *in* "The Carbohydrates," Vol. IB, p. 1376 (1980).
5. Mass spectrometry, by D. C. DeJongh, *in* "The Carbohydrates," Vol. IB, p. 1327 (1980).
6. Polarimetry, by R. J. Ferrier, *in* "The Carbohydrates," Vol. IB, p. 1354 (1980).
7. X-ray and neutron deffraction, by R. J. Ferrier, *in* "The Carbohydrates," Vol. IB, p. 1437 (1980).
8. Mutarotations and actions of acids and bases, by W. Pigman and E. F. L. J. Anet, *in* "The Carbohydrates," Vol. IA, p. 165 (1972)
9. "Carbohydrates in Solution," *Adv. Chem. Ser.* No. 117, edited by R. F. Gould. Amer. Chem. Soc., Washington D.C., 1973.
10. Carbon-13 and hydroxyl proton NMR spectra of ketoses. A conformational and compositional description of ketohexoses in solution, by A. S. Perlin, P. Herve Du Penhoat, and H. S. Isbell, *in* "Carbohydrates in Solution," *Adv. Chem. Ser.* No. 117, 39 (1973).
11. Enolization and oxidation reactions of reducing sugars, by H. S. Isbell, *in* "Carbohydrates in Solution, *Adv. Chem. Ser.* No. 117, 70 (1973).
12. Complexes of sugars with cations, by S. J. Angyal, *in* "Carbohydrates in Solution, *Adv. Chem. Ser.* No. 117, 106 (1973).
13. Conformational preferences for solvated hydroxymethyl groups in hexopyranose structures, by R. U. Lemieux and J. T. Brewer, *in* "Carbohydrates in Solution," *Adv. Chem. Ser.* No. 117, 121 (1973).
14. Conformational equilibria of acylated aldopentopyranose derivatives and favored conformations of acyclic sugar derivatives, by P. L. Durette, D. Horton, and J. D. Wander, *in* "Carbohydrates in Solution," *Adv. Chem. Ser.* No. 117, 147 (1973)
15. Conformational studies in the solid state: extrapolation to molecules in solution, by G. A. Jeffrey, *in* "Carbohydrates in Solution," *Adv. Chem. Ser.* No. 117, 177 (1973).
16. Stability constants of some carbohydrate and related complexes by potentiometric titration, by Rex Montgomery, *in* "Carbohydrates in Solution," *Adv. Chem. Ser.* No. 117, 197 (1973).
17. The Chemistry of Sugars in Boric Acid Solutions, by Terry E. Acree, *in* "Carbohydrates in Solution," *Adv. Chem. Ser.* No. 117, 208 (1973).

Chapter 5: Reactions of Monosaccharides

1. Hydrazine derivatives and related compounds, by L. Mester and H. S. El Khadem, *in* "The Carbohydrates," Vol. IB, p. 929 (1980).
2. The reaction of ammonia with acyl esters of carbohydrates, by M. E. Gelpi and R. A. Cadenas, *Adv. Carbohydr. Chem. Biochem.* **31,** 81 (1975).
3. Formation and conversion of phenylhydrazones and osazones of carbohydrates, by H. Simon and A. Kraus, *ACS Symp. Ser.* No. 39, 188 (1976).
4. Esters, by M. L. Wolfrom and W. A. Szarek, *in* "The Carbohydrates," Vol. IA, p. 217 (1972).
5. Halogen derivatives, by M. L. Wolfrom and W. A. Szarek, *in* "The Carbohydrates," Vol. IA, p. 239 (1972).
6. Glycosides, by W. G. Overend, *in* "The Carbohydrates," Vol. IA, p. 279 (1972).
7. Acyclic derivatives, by M. L. Wolfrom, *in* "The Carbohydrates," Vol. IA, p. 355 (1972).
8. Cyclic acetal derivatives of sugars and alditols, by A. B. Foster, *in* "The Carbohydrates," Vol. IA, p. 391 (1972).
9. Ethers of sugars, by J. K. N. Jones and G. W. Hay, *in* "The Carbohydrates," Vol. IA, p. 403 (1972).
10. Glycosans and anhydro sugars, by R. D. Guthrie, *in* "The Carbohydrates," Vol. IA, p. 423 (1972).
11. Alditols and derivatives, by J. S. Brimacombe and J. M. Webber, *in* "The Carbohydrates," Vol. IA, p. 479 (1972).
12. The cyclitols, by L. Anderson, *in* "The Carbohydrates," Vol. IA, p. 520 (1972).
13. Synthesis of monosaccharides, by L. Hough and A. C. Richardson, *in* "The Carbohydrates," Vol. IA, p. 114 (1972).
14. Amino sugars, by D. Horton and J. D. Wander, *in* "The Carbohydrates," Vol. IB, p. 644 (1980).
15. Deoxy and branched-chain sugars, by N. R. Williams and J. D. Wander, *in* "The Carbohydrates," Vol. IB, p. 761 (1980).
16. Thio sugars and derivatives, by D. Horton and J. D. Wander, *in* "The Carbohydrates," Vol. IB, p. 799 (1980).
17. Unsaturated sugars, by R. J. Ferrier, *in* "The Carbohydrates," Vol. IB, p. 843 (1980).
18. Glycosylamines, by H. Paulsen and K. W. Pflughaupt, *in* "The Carbohydrates," Vol. IB, p. 881 (1980).
19. Reduction of carbohydrates, by J. W. Green, *in* "The Carbohydrates," Vol. IB, p. 989 (1980).
20. Acids and other oxidation products, by O. Theander, *in* "The Carbohydrates," Vol. IB, p. 1013 (1980).
21. Oxidative reactions and degradations, by J. W. Green, *in* "The Carbohydrates," Vol. IB, p. 1101 (1980).
22. Glycol-cleavage oxidation, by A. S. Perlin, *in* "The Carbohydrates," Vol. IB, p. 1167 (1980).
23. The effects of radiation on carbohydrates, by G. O. Phillips, *in* "The Carbohydrates," Vol. IB, p. 1217 (1980).
24. Nucleosides, by C. A. Dekker and L. Goodman, *in* "The Carbohydrates," Vol. IIA, p. 1 (1970).
25. Antibiotics containing sugars, by S. Hanessian and T. H. Haskell, *in* "The Carbohydrates," Vol. IIA, p. 139 (1970).

26. Complex glycosides, by J. E. Courtois and F. Percheron, *in* "The Carbohydrates," Vol. IIA, p. 213 (1970).
27. Structures and syntheses of aminoglycoside antibiotics, by S. Umezawa, *Adv. Carbohydr. Chem. Biochem.* **30,** 111 (1974).
28. Deamination of carbohydrate amines and related compounds, by J. M. Williams, *Adv. Carbohydr. Chem. Biochem.* **31,** 9 (1975).
29. Dithioacetals of sugars, by J. D. Wander and D. Horton, *Adv. Carbohydr. Chem. Biochem.* **32,** 16 (1976).
30. Relative reactivities of hydroxyl groups in carbohydrates, by A. H. Haines, *Adv. Carbohydr. Chem. Biochem.* **33,** 11 (1976).
31. Synthesis of naturally occurring *C*-nucleosides, their analogs, and functionalized *C*-glycosyl precursors, by S. Hanessian and A. G. Pernet, *Adv. Carbohydr. Chem. Biochem.* **33,** 111 (1976).
32. Reactions of D-glucofuranurono-6,3-lactone, by K. Dax and H. Weidmann, *Adv. Carbohydr. Chem. Biochem.* **33,** 189 (1976).
33. 1,6-Anhydro derivatives of aldohexoses, by M. Černý and J. Staněk, Jr., *Adv. Carbohydr. Chem. Biochem.* **34,** 23 (1977).
34. Cyclic acetals of the aldoses and aldosides, by A. N. De Belder, *Adv. Carbohydr. Chem. Biochem.* **34,** 179 (1977).
35. The Koenigs–Knorr reaction, by K. Igarashi, *Adv. Carbohydr. Chem. Biochem.* **34,** 243 (1977).
36. Application of ethylboron compounds in carbohydrate chemistry, by R. Koster and W. V. Dahlhoff, *ACS Symp. Ser.* No. 39, 1 (1976).
37. Carbohydrate boronates, by R. J. Ferrier, *Adv. Carbohydr. Chem. Biochem.* **35,** 31 (1978).
38. Glycosiduronic acids and related compounds, by D. Keglević, *Adv. Carbohydr. Chem. Biochem.* **36,** 57 (1979).
39. Free-radical reactions of carbohydrates as studied by radiation techniques, by C. von Sonntag, *Adv. Carbohydr. Chem. Biochem.* **37,** 7 (1980).
40. Synthesis of L-ascorbic acid, by T. C. Crawford and S. A. Crawford, *Adv. Carbohydr. Chem. Biochem.* **37,** 79 (1980).
41. Photochemical reactions of carbohydrates, by R. W. Binkley, *Adv. Carbohydr. Chem. Biochem.* **38,** 105 (1981).
42. Fluorinated carbohydrates, by A. E. Penglis, *Adv. Carbohydr. Chem. Biochem.* **38,** 195 (1981).
43. The selective removal of protecting groups in carbohydrate chemistry, by A. H. Haines, *Adv. Carbohydr. Chem. Biochem.* **39,** 13 (1981).
44. Synthesis and polymerization of anhydro sugars, by C. Schuerch, *Adv. Carbohydr. Chem. Biochem.* **39,** 157 (1981).
45. The synthesis of sugars from non-carbohydrate substrates, by A. Zamojski, A. Banaszek, and G. Grynkiewicz, *Adv. Carbohydr. Chem. Biochem.* **40,** 1 (1982).
46. Synthesis of branched-chain sugars, by J. Yoshimura, *Adv. Carbohydr. Chem. Biochem.* **42,** 69 (1984).
47. Sugar analogs having phosphorus in the hemiacetal ring, by H. Yamamoto and S. Inokawa, *Adv. Carbohydr. Chem. Biochem.* **42,** 135 (1984).
48. Phosphates and other inorganic esters, by D. L. MacDonald, *in* "The Carbohydrates," Vol. IA., p. 253 (1972).
49. Applications of gas–liquid chromatography to carbohydrates, by G. G. S. Dutton, *Adv. Carbohydr. Chem. Biochem.* **28,** 11 (1973); **30,** 9 (1974).

50. F.a.b.-Mass spectrometry of carbohydrates, by A. Dell, *Adv. Carbohydr. Chem. Biochem.* **45,** 19 (1987).
51. Separation methods: Chromatography and electrophoresis, by M. I. Horowitz, *in* "The Carbohydrates," Vol. IB, p. 1445 (1980).

Chapter 6: Structure of Oligosaccharides

1. The metabolism of α,α-trehalose, by A. D. Elbein, *Adv. Carbohydr. Chem. Biochem.* **30,** 227 (1974).
2. Chemistry and biochemistry of apiose, by R. R. Watson and N. S. Orenstein, *Adv. Carbohydr. Chem. Biochem.* **31,** 135 (1975).
3. The lectins: Carbohydrate-binding proteins of plants and animals, by I. J. Goldstein and C. E. Hayes, *Adv. Carbohydr. Chem. Biochem.* **35,** 127 (1978).
4. The chemistry of maltose, by R. Khan, *Adv. Carbohydr. Chem. Biochem.* **39,** 213 (1981).
5. The utilization of disaccharides and some other sugars by yeasts, by J. A. Barnett, *Adv. Carbohydr. Chem. Biochem.* **39,** 347 (1981).
6. Chemistry, metabolism, and biological functions of sialic acids, by R. Schauer, *Adv. Carbohydr. Chem. Biochem.* **40,** 131 (1982).
7. High-resolution, ^{1}H-nuclear magnetic resonance spectroscopy as a tool in the structural analysis of carbohydrates related to glycoproteins, by J. F. G. Vliegenthart, L. Dorland, and H. van Halbeek, *Adv. Carbohydr. Chem. Biochem.* **41,** 209 (1983).
8. Carbon-13 nuclear magnetic resonance data for oligosaccharides, by K. Bock, C. Pedersen, and H. Pedersen, *Adv. Carbohydr. Chem. Biochem.* **42,** 193 (1984).

Chapter 7: Chemical Synthesis and Modifications of Oligosaccharides, Nucleotides, and Aminoglycoside Antibiotics

1. Structures and syntheses of aminoglycoside antibiotics, by S. Umezawa, *Adv. Carbohydr. Chem. Biochem.* **30,** 111 (1974).
2. Biochemical mechanism of resistance to aminoglycosidic antibiotics, by H. Umezawa, *Adv. Carbohydr. Chem. Biochem.* **30,** 183 (1974).
3. The synthesis of polynucleotides, by M. Ikehara, E. Ohtsuka, and A. F. Markham, *Adv. Carbohydr. Chem. Biochem.* **36,** 135 (1979).
4. Biosynthesis and catabolism of glycosphingolipids, by Y. Li and S. Li, *Adv. Carbohydr. Chem. Biochem.* **40,** 235 (1982).
5. "The Total Synthesis of Natural Products," edited by J. ApSimon, Vol. 2. Wiley, New York, 1976.

Appendix: Answers to Problems

Chapter 3

1. D-erythronic acid
2. L-*arabino*-hexosulose
3. D-*erythro*-3-pentulosonic acid
4. *erythro*-3,4-hexodiulose
5. D-*glycero*-D-*manno*-heptose
6. D-glucose diethyl dithioacetal
7. D-fructose diethyl acetal
8. 2-(D-*arabino*-tetrahydroxybutyl)furan
9. phenyl β-D-xylopyranoside
10. 3-deoxy-β-D-*erythro*-pentose
11. 2,5-anhydro-D-mannitol
12. 2-amino-2-deoxy-α-D-glucopyranose in the $^4C_1$ conformation

Chapter 4

1. (a) 3
   (b) 1
   (c) 2

2. (a) $^1C_4$
   (b) $^1C_4$
   (c) $^4C_1$

3. all β $^4C_1$

4. (a) inositol
(b) glucitol
(c) mannitol
Note how symmetry reduces the number of signals.

5. (a) protonated base (B + H)
(b) B—CH—OH (B + 30)
(c) B—CH_2—CHOH (B + 44)
(d) M − CH_2O (M − 30)
(e) M^+

6. Since the reaction is first order, only two structures, namely the α and β pyranoses, contribute significantly to the equilibrium. The remaining structures exist in too small concentrations to affect the measurements.

Chapter 5

1. (a) arabinose + ^{13}C-labeled KCN, followed by hydrolysis, lactonization, and reduction
(b) hydrolyze the unlabeled nitrile with ^{18}O-labeled water
(c) reduce D-glucose with $LiAlD_4$; then oxidize, first to sorbose and then to ascorbic acid
(d) di-*O*-isopropylideneglucose is tosylated, then treated with labeled hydrazine, and the product reduced.

3. (a) Glycosidate with MeOH/HCl; treat with PhCHO to form the benzylidene ring; tosylate.
(b) Add NaOMe/MeOH to form the epoxide ring.
(c) Add MeMgI/ether to methylate and open the epoxide ring (axial OH formed).
(d) Oxidize the OH to C=O with DCC/Me_2SO; reduce the C=O with LAH (attack from the least hindered side) to form an OH down.
(e) Oxidize the OH to C=O as in d; isomerization with base to bring the Me up; reduce with LAH; methylate with MeI/NaH.
(f) Open the benzylidene ring with NBS; reduce the bromide with LAH.

Chapter 6

1. (a) 52.6%
(b) 2
(c) D-arabinonic acid and galactose
(d) 1,3,6-tri-*O*-methyl-D-fructofuranose and 2,3,4,6-tetra-*O*-methyl-D-galactopyranose

2. (a) A–B–A–C–A, B–A–C–A–B, A–C–A–B–A, C–A–B–A–C
(b) A–B–A–C, B–A–C–A, A–C–A–B, C–A–B–A
(c) A–B–A, B–A–C, A–C–A, C–A–B
(d) A–B, B–A, A–C, C–A

3. I. 1 = a; 2 = c; 3 = Me of a.
II. 1 = b (α); 2 = a (α); 3 = b (α); 4 = b (β); 5 = a (β); 6 = Me of a.
III. 1, 2 = a; 3, 4 = c; 5, 8 = CH_2 of e; 6, 7 = Me of a, e.
IV. 1, 4 = a; 2 = b (α); 3 = b (β); 5, 6 = Me of a.

4. See Table I.

5. Ester was attached on position 3.

Chapter 7

1. Form the benzyl glycoside of *N*-acetylglucosamine, tritylate, acetylate, and detritylate. React this with tri-*O*-benzylfucopyranosyl bromide under Koenigs–Knorr conditions. Finally, deacetylate, and hydrogenate to remove the benzyl groups.

2.

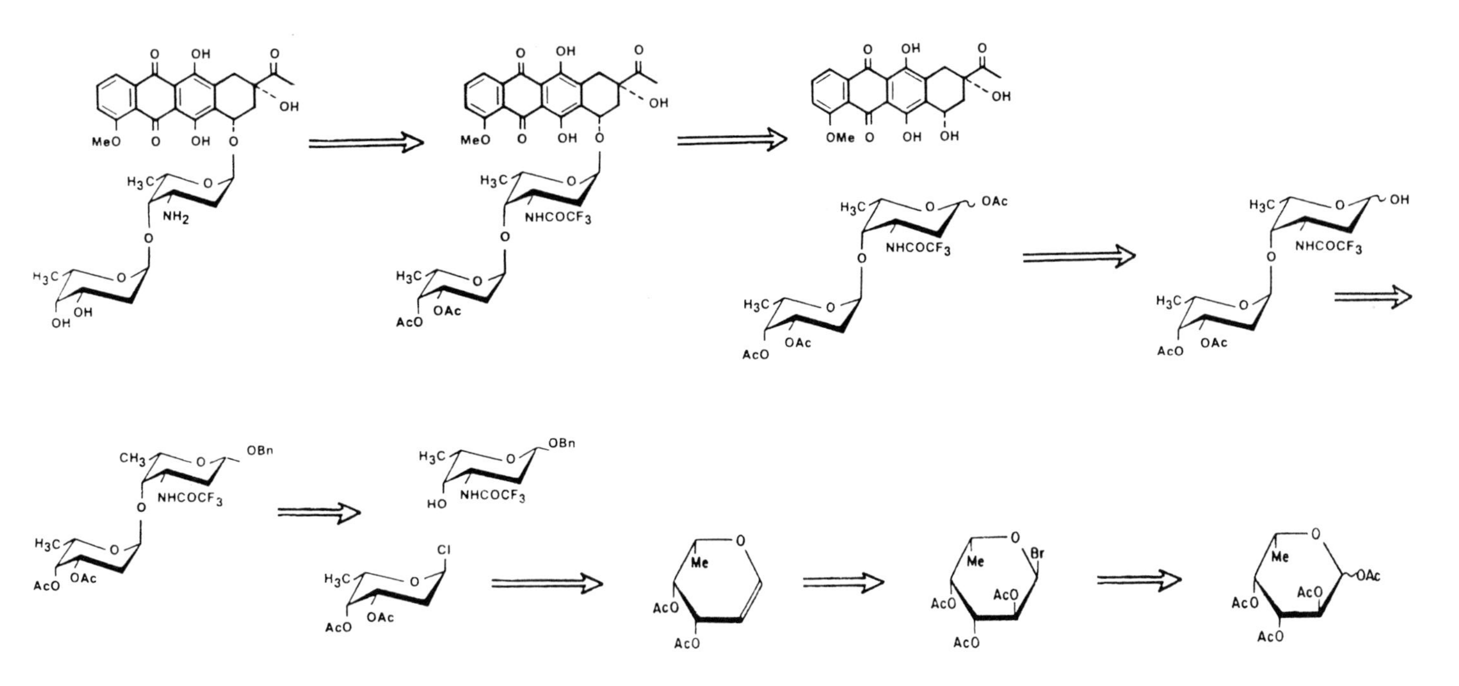

3.

$$\text{(5'-O-R)thymidine} + \text{3'-O-acetylthymidine 5'-phosphate} \xrightarrow[\text{2) } OH^- \text{; 3) } H_3PO_4]{\text{1) DCC}} \text{RO-}CH_2\text{-dT-3'-O-P(O)(OH)-O-5'-}CH_2\text{-dT-3'-O-P(O)(OH)}_2$$

Index

K

L

M

N

O

P

R